Sentinels of the Sun

Forecasting Space Weather

Barbara B. Poppe
with Kristen P. Jorden

Johnson Books
Boulder

Dedicated to our mothers

Published by Johnson Books, a division of Big Earth Publishing, 3005 Center Green Drive, Suite 220, Boulder, Colorado 80301. Visit our website at www.bigearthpublishing.com
E-mail: books@bigearthpublishing.com

Front Cover: From an illustration by Miles Yergenson
Back cover: An image from Transition Region and Coronal Explorer, TRACE, Stanford-Lockheed Institute for Space Research and the NASA Small Explorer program.
Cover design: Polly Christensen
Text design and composition: Michel Reynolds
Figure 7.11, page 149, magazine permission by The Racing Pigeon Publishing Co Ltd, email rp@racing pigeon.freeserve.co.uk; and the American Racing Pigeon Union, Oklahoma City, OK, www.pigeon.org.

9 8 7 6 5 4 3 2 1

Library of Congress Cataloging-in-Publication Data
Poppe, Barbara B.
Sentinels of the sun: forecasting space weather / Barbara B. Poppe, Kristen P. Jorden.
p. cm.
Includes bibliographical references and index.
ISBN 1-55566-379-6
1. Space environment. 2. Solar activity. I. Jorden, Kristen P. II. Title.
QB505.P67 2006
551.51′4--dc22

2006005284

Printed in the United States of America

Contents

Preface

The field of space weather has come to an age in which those who gave birth to it—those who came to understand the Sun and warn others about it—are at the end of their careers. Younger scientists will advance the field, uncovering new secrets of the Sun and creating better services, but they will not carry the history of their field in their hearts, as the original parents have.

This book is a tribute to those parents and pioneers who will not be able to relate the history themselves. Having worked at the Space Environment Center (SEC) for twenty-one years, I feel very much immersed in this history and am pleased to be a voice for those who have lived it. It is a history that I wanted to see recorded and remembered as much as I wanted to tell interested readers about this fascinating field.

I am deeply indebted to Gary Heckman, who shared a respect for this history and contributed the first bits and pieces before he died, and to Joe Hirman, who persevered to see the project through. Both men were dedicated individuals who lived this history for over thirty-five years. As with any undertaking of this size, it is impossible to name all the people who helped make this book possible. Past and present SEC staff helped enormously in patiently explaining the science and clarifying the history, right down to the minute details. Individuals in the field, including many from NOAA, NASA, USAF, HAO, universities, and research organizations, had much to add to the book. I hope they all enjoy reading about their histories and see themselves as part of the story, whether named or not. The users of space weather will also find themselves on these pages due to their invaluable contributions.

I would like to thank Meg Austin of the University Center for Atmospheric Research for shepherding the project with enthusiasm and encouragement, and the folks at Johnson Books who took such an interest in this book. Finally, I thank my devoted family for their support.

Barbara B. Poppe

In collaborating on this book, I had two visions: to introduce the past to future scientists, and to give a gift to present readers. As a history teacher, I am constantly impressed that many people do not regard the 1980s as history—to them, history is "a long time ago." Although that point can be argued, the last few decades will soon become history: the baby-boom generation will pass on and the memories its members cherish will fade away. Forward thinkers realize the value of capturing history while it is in the making, while those who lived it are still around to share silly and poignant stories. Future generations deserve to hear history from the mouths of the people who witnessed it.

Far from being merely a dry chronicle of past events, the book is intended to provide something rare to our contemporaries. Too often scientists (or any academics, for that matter) write for their own community, forgetting that many lay readers have an insatiable curiosity to know the world around them. Some academics (notably in the field of history) are taking great pains to dejargonize their writing and share their passion with the lay reader. With that goal in mind, I took the sometimes inaccessible technical writing and attempted to transform it into something ultimately readable.

Many thanks to my husband, Noel, for his patient ear, scientific eye, and endless words of encouragement.

Kristen P. Jorden

A Chronology
of Significant Events Resulting from Solar Activity

August–September 1859 A huge auroral event, following a solar flare visible to the naked eye, disabled much of the world's telegraph network. Innumer-able reports, ships' logs, and newspapers around the world recorded aurora sightings during this event, even one as far south as central India, only 18 degrees above the equator.

October 1935 Radio disturbances, known as Shortwave Fade, were first correlated with major solar flares when radio communications failed within minutes of a large flare's occurrence on the sunlit side of Earth.

March 1940 The first reported incidences of power grid problems due to solar activity occurred all across the Northeast, from Pennsylvania through New England, and in Minnesota, Ontario, and Quebec. At the same time, nearly all overseas radio telephones and long-distance land telephones failed for a number of days.

September 1941 The greatest display of aurora ever seen in Washington, D.C., might have been coincident with magnetic disturbances that caused radio communications and compasses to fail on bombers trying to navigate back to England. This temporary communications loss leading to bombers lost at sea or crashing on land is a well-known tale, though not well documented.

February 1958 The city of Toronto experienced a short power blackout. Solar activity interfered with the new TAT-1 transatlantic communication cables as well.

August 1972 Solar activity affected power grids across the Northern Hemi-sphere and compromised navigation on the St. Lawrence River. Measurements of the radiation hazard for astronauts were so high that the National Aeronautics and Space Administration (NASA) had to plan for more safety measures on the upcoming Apollo 17 manned mission to the Moon.

July 1982 Railroad traffic signals in Sweden automatically switched to red (stop) as a result of induced currents from a geomagnetic storm.

March 1989 Energy provider Hydro-Quebec failed and caused a nine-hour blackout over the entire province of Quebec. The main network failed within a minute of receiving intense storm-caused voltage, and the rest of the grid followed, collapsing piece by piece in twenty-five seconds. Power grids failed and caused transformer damage throughout the northeastern United States.

September 1989 The supersonic transport commonly known as the Concorde registered a radiation alert during a solar radiation storm while on a flight from France to the United States. This sobering event illustrated the increasing radiation risk to humans as they began to fly higher in latitude and altitude.

January 1994 After days of minor space weather storms, the Canadian telecommunications Anik satellites failed, disrupting commercial exchange of data, long-distance telephone service throughout Canada, and TV broadcasts to 450 affiliates and the cable network.

February 1994 Broadcasts of the winter Olympic games, held in Norway, were disrupted for fifty minutes when the Japanese BS-3a TV satellite was damaged by space weather. Consequently, many Japanese missed out on some of their team's competitions, a great disappointment.

July 1998–December 2003 The Japanese satellite Nozomi fell prey to several space weather storms, damaging communications, power systems, and power cells. After many mishaps, all of which delayed the planned mission to Mars, the mission was abandoned four years after the satellite was supposed to have reached Mars.

July 2000 Geosynchronous communication satellites and other spacecraft became disoriented or suffered other damage from a significant storm. The star tracker on NASA's Stardust spacecraft lost its orientation but was later recovered. The Japanese low-Earth-orbit science mission experienced increased drag, consequently losing the altitude necessary to keep it oriented and prematurely ending its mission life. The NASA/European Space Agency (ESA) Solar and Heliospheric Observatory suffered permanent solar-panel output degradation; in other words, it lost valuable solar power.

October–November 2003 A massive series of storms jeopardized all spacecraft in orbit. Some operators took quick precautions to protect their equipment, but some satellites were lost. Severe communication problems and heightened radiation risks kept airlines out of the polar regions for much of three days. The last storm of the series included an X28 flare, by far the largest ever recorded, but luckily the emissions blasted into space away from Earth and so caused little damage.

January 2005 High radiation risk and poor communications from solar storms forced airlines to reroute planes away from high latitudes, causing flights to be longer and more expensive. At least one airline lost $1 million because of the two-day storm. Astronauts in the International Space Station took cover in protected parts of the spacecraft to avoid radiation damage. Many satellites lost data by shutting down to ride out the storm.

Acronym List

ACE	Advanced Composition Explorer
AFWA	Air Force Weather Agency
ARINC	Aeronautical Radio, Inc.
ARPA	Advanced Research Projects Agency
ARRL	American Radio Relay League
ATM	Apollo Telescope Mount
CME	coronal mass ejection
COSPAR	Committee on Space Research
CRADA	Cooperative Research and Development Agreement
CRPL	Central Radio Propagation Laboratory
CYI	call letters for the Canary Islands SPAN station
DARPA	Defense Advanced Research Projects Agency
DGPS	Differential Global Positioning System
DoC	Department of Commerce
DoD	Department of Defense
DRTE	Defence Research Telecommunications Establishment
DSRC	David Skaggs Research Center
ESA	European Space Agency
ESSA	Environmental Science Service Agency
FAA	Federal Aviation Administration
FAGS	Federation of Astronomical and Geophysical Data Analysis Services
FY	fiscal year
GIC	geomagnetic-induced current
GOES	Geostationary Operational Environmental Satellite
GPS	Global Positioning System
GSA	General Services Administration
HANDS	High Altitude Nuclear Detection Studies
HAO	High Altitude Observatory
HF	high frequency
ICSU	International Council for Science
IGY	International Geophysical Year
IPY	International Polar Year
IRPL	Interservice Radio Propagation Laboratory
ISEE	International Sun Earth Explorer
ISES	International Space Environment Service
ISIB	InterService Ionosphere Bureau
IUGG	International Union of Geodesy and Geophysics
IUWDS	International URSIgram and World Days Service

LASCO	Large Angle and Spectrometric Coronagraph Experiment
Loran	long-range navigation
LUF	lowest usable frequency
MUF	maximum usable frequency
NASA	National Aeronautics and Space Administration
NASDA	(Japan's) National Space Development Agency
NATO	North Atlantic Treaty Organization
NBS	National Bureau of Standards
NESDIS	National Environmental Satellite, Data, and Information Services
NGS	National Geodetic Survey
NIST	National Institute of Standards and Technology
NOAA	National Oceanic and Atmospheric Administration
NORAD	North American Aerospace Defense Command
NPOESS	National Polar-Orbiting Operational Environmental Satellite System
NRL	Naval Research Laboratory
NSWP	National Space Weather Program
NWS	National Weather Service
OAR	(NOAA's Office of) Oceanic and Atmospheric Research
RTSW	real-time solar wind
Satcom	satellite communications
SCOSTEP	Scientific Committee on Solar-Terrestrial Physics
SDFC	Space Disturbances Forecast Center
SDL	Space Disturbances Laboratory
SEC	Space Environment Center
SEL	Space Environment Laboratory
SEON	Solar Electro-Optical Network
SESC	Space Environment Services Center
SEU	single-event upset
SOHO	Solar and Heliospheric Observatory
SPAN	Solar Particle Alert Network
SRAG	Solar Radiation Analysis Group
STEREO	Solar Terrestrial Relations Observatory
SXI	solar X-ray imager
TIROS	Television InfraRed Observation Satellite
URSI	Union Radio–Scientifique Internationale
USAF	U.S. Air Force
WAAS	Wide Area Augmentation System
WDC	World Data Centers
WMO	World Meteorological Organization

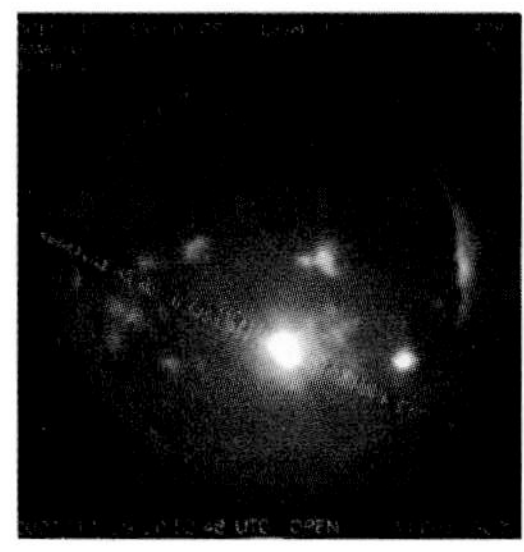

Introduction

Joe Kunches kicked back in his chair, snapped his pen on and off, and wished for another cup of coffee. The computer monitor in front of him showed nothing remarkable, just a series of pictures of a quiet Sun as it turned through its waning solar cycle. The pictures were familiar, and the small number of sunspots had the whole Sun to themselves at this time in the solar cycle. It was October 21, 2003, and Joe was "on the desk" at the Forecast Center at the Space Environment Center (SEC), part of the National Oceanic and Atmospheric Administration (NOAA) in Boulder, Colorado.

Unimpressed with the monitor in front of him, he turned to the computer to his left and scrolled through his e-mail, absentmindedly deleting the junk and reminding himself to answer the important messages later.

Joe had been watching the Sun for years. As a young man he had become a NOAA Corps officer and been assigned to be a solar forecaster. After working at SEC for a short period, he took some time away from space weather to get a taste of the world but eventually returned to SEC as a civil employee. Now, thirty years later, he was the senior forecaster and head of the Forecast and Analysis Branch. As the leader of the forecasting and operation specialist team, he was appreciated for his youthful charm and humor and respected for his vast experience in the business.

Earlier in the day a group of fifth graders had come through the Forecast Center, mostly wanting to play with the knobs and buttons, but the teacher had insisted that they learn something about the Sun (Fig. 0.1). Joe had explained that there was something called a solar cycle, a roughly eleven-year cycle of solar activity that runs from a minimum number of sunspots to a maximum, and back to a minimum. After Solar Maximum, the Sun can settle down to a bland and quiet state. This day, three years after the wild times of Solar Maximum, the early-October sunspot count reflected another quiet month in the unremarkable waning phase of an average solar cycle. Joe could see that the students were hoping for something more exciting than "a quiet month in the unremarkable waning phase. . . ." He had told them that in the past sixty days his department had issued two small alerts, warning high-flying aircraft, companies with satellites orbiting above the Earth, and radio users that the Sun was producing more damaging emissions than normal. The students had perked up at the thought of damaging radiation.

Figure 0.1: *Tours for high school students showcase the SEC Forecast Center in Boulder, Colorado. Larry Combs, along with several other forecasters, gives many tours in a year. (NOAA)*

By now all the kids had left, and Joe rifled through papers in the quiet of the forecast room. He had an inkling that the excitement the kids wanted might very well be on the way. Looking with satellite instruments at the west edge of the Sun two days before, he had seen a coronal mass ejection from a solar flare on the *back* side of the Sun; he could tell something was coming. Now, toward the end of the day, out of the corner of his eye, he saw a series of pictures on the monitors that made him pause. The constantly rotating Sun now revealed a cluster of large, dark spots. They were only slightly larger than the last time he had seen them, but he was experienced enough to pick out even the slightest change. The areas around the dark cluster of spots and the bright "active regions" had begun to come to a boil.

Calmly, but with excitement in his heart, Joe greeted the next person who walked in, this time the director of the center. "I think you'd better come and take a look at this," Joe urged. Before they could finish speculating about this new development, the small forecast room was packed with forecasters and scientists, one group jabbering about the implications for the forecast and the other vying for a look at the images on the monitors.

Over the next week the sunspots grew in size. They were no longer distinct, isolated spots but were intermixed with the active regions. The watch in the Forecast Center continued day after day as the likelihood of a solar storm grew. One week later, on October 28, it began (Fig. 0.2).

Joe was joined by the full rotation of forecasters, who often doubled up for duty to handle the workload. This was proving to be the most

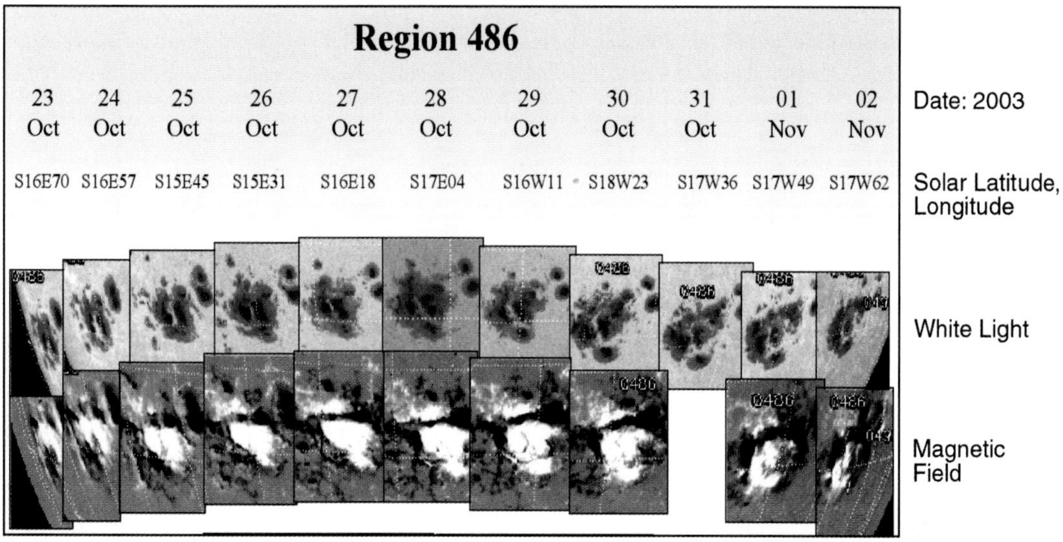

Figure 0.2: *The solar disk transit of Super Region 486 over eleven days is shown in two ways: a full-spectrum white-light image and a magnetic field image. The storm from this region that began on October 28, 2003, affected Earth's magnetic field and produced a G5 (extreme) geomagnetic storm. The storm lasted for twenty-seven hours and was the sixth most intense geomagnetic storm on record. (Big Bear Solar Observatory, New Jersey Institute of Technology)*

active and demanding solar activity in years. NOAA forecasters had three tasks: keep a close eye on the areas of solar activity and analyze the likelihood of storms; issue alerts, warnings, watches, and forecasts with the correct urgency and timeliness; and answer questions.

Over the next twenty days, from late October to early November, NOAA forecasters issued over 250 watches, warnings, and alerts, the result of explosions from three of the largest sunspot clusters in over ten years. These warnings served to alert businesses that would be affected by the increase in particle radiation, x-ray emissions, and the solar wind (Fig. 0.3). Because businesses tend to be reluctant to lay out their troubles to the public, they often fail to let SEC know what troubles they have experienced. This time, however, reports from affected businesses came pouring in from around the world. The Sydkraft power utility group in Sweden reported that strong geomagnetic-induced currents (GICs) over northern Europe had caused transformer problems and even a system failure and subsequent blackout. Radiation storm levels were high enough to prompt NASA flight specialists to issue a directive to the International Space Station astronauts to take precautionary shelter. NASA also decided to do a power-down of the $1 billion robotic arm and workstation, which are sensitive to radiation events.

Airlines took unprecedented actions with their high-latitude routes to avoid the high radiation levels and communication blackout areas. Rerouted flights cost airlines $10,000 to $100,000 per flight. Numerous anomalies (a general term for glitches that might be easily fixed or catastrophic) were reported by deep-space missions and by satellites at all orbits. NASA Space Science Mission Operations indicated that approximately 59 percent of the Earth and space science satellite missions

Figure 0.3: *Space weather disturbances affect space shuttles, satellites, airplanes, ships, power grids, pipelines, radio communications, and even migrating birds in flight. (NOAA)*

reported anomalies: in at least half the missions, the satellite was shut down and put into "safe mode." The instruments on the Mars Odyssey orbiter, the first to measure the solar radiation around another planet, stopped working permanently, and are now considered a total loss. The storms are suspected to have caused the total loss of communications with the $640 million ADEOS-2 spacecraft. On board the ADEOS-2 was the $150 million NASA SeaWinds instrument. Both were dreams come true for the scientists and engineers who worked on them. Both are floating space junk now.

Questions from the media were answered as quickly as possible. Several forecasters and researchers fielded questions from their offices, away from the frantic activity in the Forecast Center. The widespread concerns of the affected commercial sectors generated intense global media interest.

Solar images and stories of solar activity flashed across major newspapers around the world, making "solar flares" a household term. SEC staff participated in over three hundred news broadcasts and interviews, assisting media outlets from Chile to Hong Kong. The high levels of activity fueled more public and media interest than any other solar event to date. Customer concerns and heightened public awareness produced a frenzy of interest in the SEC web page, which saw the daily average hit rate of 500,000 rise to over nineteen million hits on October 29 (Fig. 0.4). A hasty transfer to a NOAA web farm, usually used for large hurricane updates, made this unusual traffic possible.

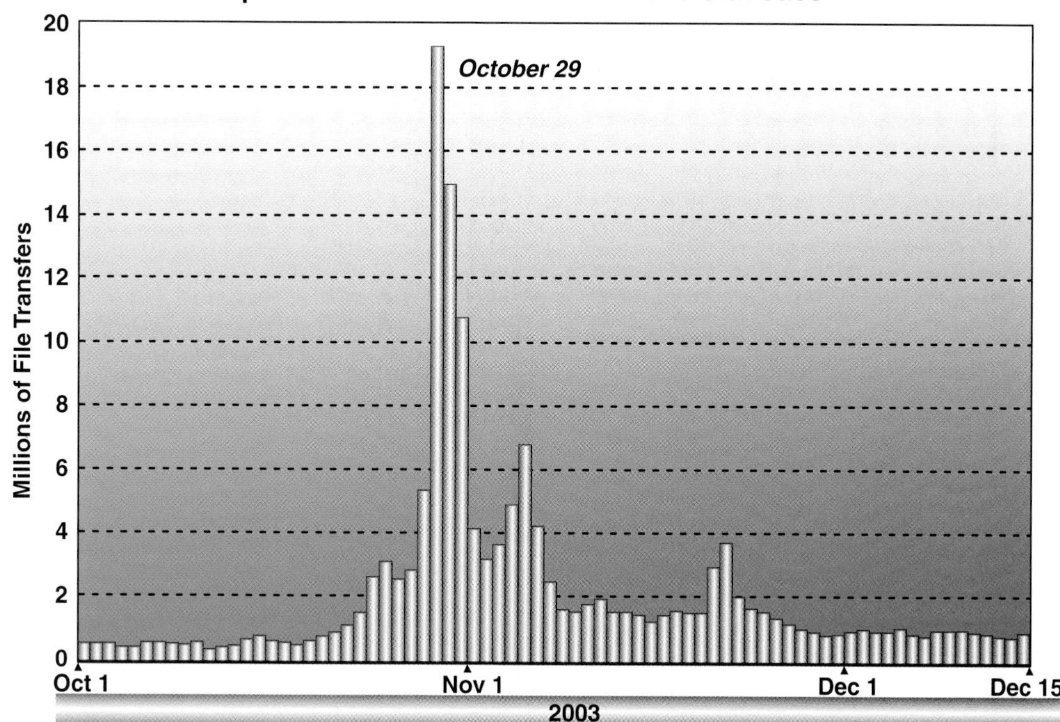

Figure 0.4: *SEC's website received nineteen million hits in twenty-four hours from anxious operators and excited scientists during the 2003 storm. (NOAA)*

The Halloween Storms, as they came to be called, occurred late in Solar Cycle 23, when it is unusual to see this extreme level of activity. In the twenty days of activity between October 19 and November 5, SEC staff witnessed a total of seventeen major flares (one was a severe radio blackout storm—rated R4 out of 5), six radiation storms (one was a severe solar radiation storm—rated S4 out of 5), and four severe geomagnetic storms (two of which reached the extreme level—rated G5 out of 5) (Figs. 0.5 and 0.6).

This activity included the most intense flare ever measured by a NOAA Geostationary Operational Environmental Satellite instrument—a

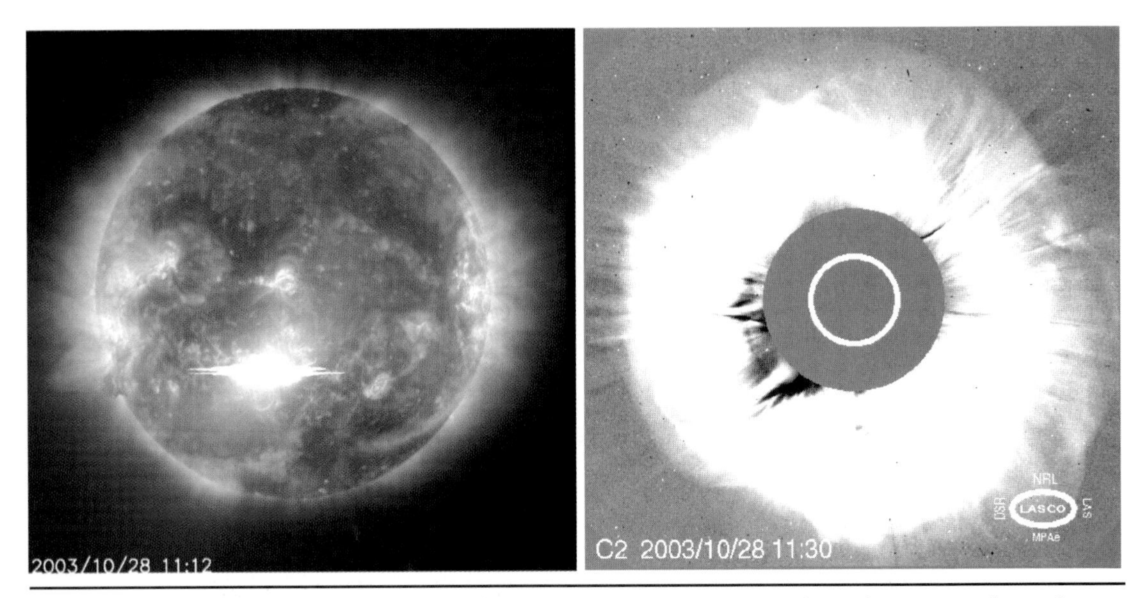

Figure 0.5: *(Left) The huge X17 flare on October 28, shown here as an image from the extreme ultraviolet imaging telescope on the Solar and Heliospheric Observatory (SOHO) satellite, caused an R4 radio blackout storm. (Right) This image, taken by the Large Angle and Spectrometric Coronagraph Experiment (LASCO) on SOHO, shows the full halo CME on October 28. The gray disk in the center blocks out the very bright light of the Sun, which is indicated by the white ring in the center of the disk. The bright emission fully surrounding the Sun (the halo) shows that the CME is directed toward Earth. (NASA/ESA)*

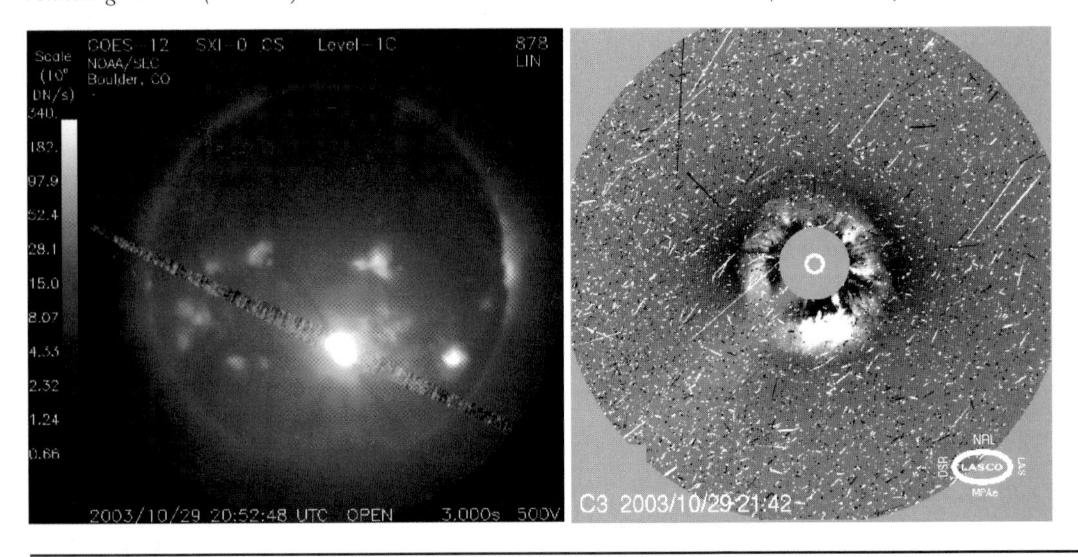

Figure 0.6: *(Left) The October 29 solar flare just to the right of and below the center of the Sun produced an S4 solar radiation storm on Earth. The image was captured by the GOES solar x-ray imager (SXI). The streak from upper left to lower right is due to refraction in the optics. (NOAA) (Right) A different view of the halo CME of October 29, as seen by the LASCO imager, shows the effect of the radiation storm even before the particles hit the Earth. The "hash" in the picture is the particles from the Sun actually hitting the imager. (NASA)*

huge X28 flare on November 4 (Fig. 0.7). No one in the business of space weather had ever seen such an enormous flare. It put out more energy during its short lifetime (several hours) than humankind has used on Earth to date. Luckily, because the huge X28 flare was directed away from Earth, the subsequent radiation storm and geomagnetic storm reached only minor and moderate levels. However, many of the other flares released very large radiation storms.

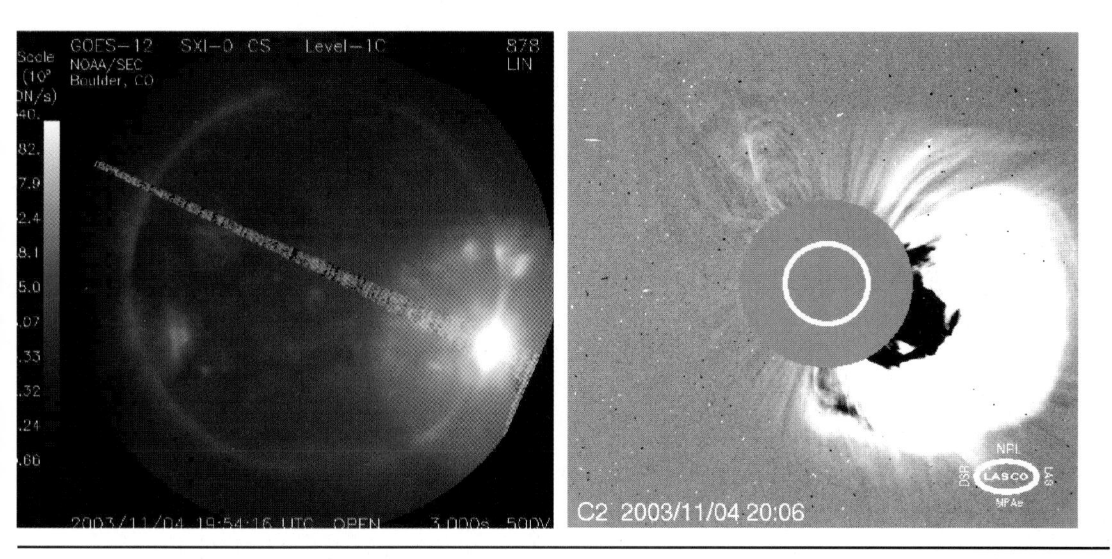

Figure 0.7: *(Left) The largest flare ever seen by the GOES instruments, estimated to be an X28, can be seen to the far right in the picture, taken by SXI on November 4. (NOAA) (Right) The LASCO imager captured the CME generated with the flare. Because of the position and direction of the event (seen on the right side), the emissions virtually missed the Earth and caused little storming. (NASA) A full report of the Halloween Storms and their effects is available at http://www.sec.noaa.gov/AboutSEC/SWstorms_assessment.pdf.*

Geomagnetic storm periods were observed on twelve of the next twenty days. The extreme and long-lasting geomagnetic storms on October 29 and 30 spread the aurora over vast regions at the middle latitudes and even low latitudes (Fig. 0.8). Auroras, typically seen only in arctic and antarctic latitudes, were seen from California to Houston to Florida. Tremendous aurora viewing was also reported from Australia, mid-Europe, and even as near to the equator as the Mediterranean countries (Fig. 0.9). The last time SEC had seen this kind of activity so late in the cycle had been in April and May of 1984.

During such an active time, SEC relied heavily on other organizations around the world that also monitor solar activity. The Regional Warning Centers of the International Space Environment Service and Space Weather Operations Center of the Air Force Weather Agency (SEC's Air Force sister center) proved most helpful. In general, daily coordination calls are standard procedure and play an important role

Figure 0.8: *On November 20, 2003, the aurora appeared over the National Center for Atmospheric Research in Boulder, Colorado. To accentuate the aurora, the building in the lower left corner is overexposed. (Stan Solomon, NCAR)*

in developing accurate and consistent forecasts of space weather. However, during the October–November activity period, the level of coordination between the centers was unprecedented. The solar observations provided by the Air Force's Solar Electro-Optical Network proved to be especially invaluable, since the solar x-ray imager on the NOAA GOES-12 satellite was not working during the first eleven major flares (Fig. 0.10). SEC used these Air Force observations of flare location, radio burst intensity, and radio sweep characteristics in its radiation storm and geomagnetic storm forecasts. From the observations, forecasters could provide information to NASA, the Federal Aviation Administration, National Weather Service Doppler radar operators, and thousands of other businesses or organizations that wanted to protect their systems or, in some cases, their health.

NASA provided coronagraph observations from its research spacecraft, the Solar and Heliospheric Observatory (SOHO), which had the only direct view of coronal mass ejections (CMEs). Halo CMEs, unless aimed directly away from Earth, hit the Earth dead-on and cause large geomagnetic disturbances. In showing CMEs, the data from SOHO gave the best and earliest predictor of strong geomagnetic disturbances.

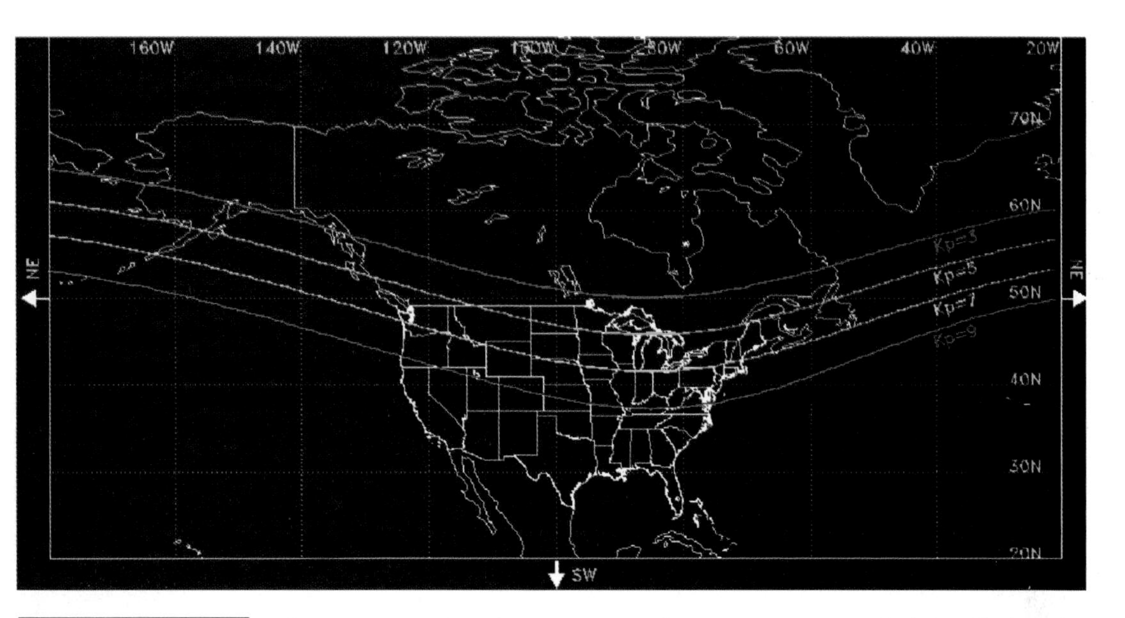

Figure 0.9: *Likelihood of seeing the aurora depends on geomagnetic latitude as shown in this plot. People at higher latitudes have more of a chance to see the spectacle than people in the lower latitudes. The October 29–30, 2003, storm was so large that people in states as low as Texas and Florida saw the aurora. (NOAA)*

The team running the coronagraph on SOHO provided detailed analysis and reporting of the many CMEs during the October–November period, greatly aiding SEC forecasters. "What would we do without SOHO?" was a common sentiment expressed by forecasters.

The dramatic activity in October–November 2003 underscored our current level of understanding of the Sun, which, until recently, has been incredibly limited. The Sun has been watched and worshipped for millennia by cultures around the world. It was rightly seen as the source of weather, heat, and light. For thousands of years the Sun was also seen as sacred, mysterious, and powerful. To the present day, the Sun has been an object of both fascination and respect.

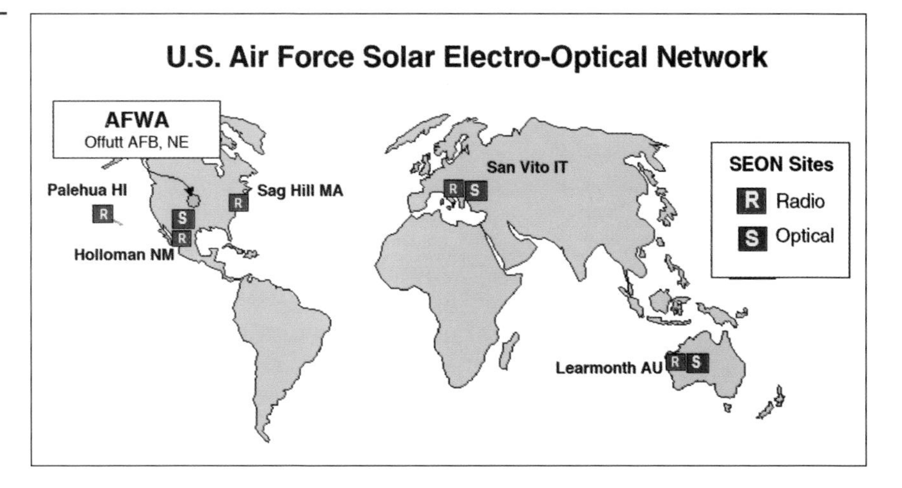

Figure 0.10: *The Solar Electro-Optical Network (SEON), a network of telescope sites that included RSTN radio telescopes and SOON optical telescopes, provided critical real-time observations during the severe October–November 2003 solar storms. (AFWA)*

Because of our technological developments in the last hundred years, we have been forced to examine the workings of the Sun. Human reliance on computers, satellites, cell phones, interconnecting power grids, and aircraft—technologies that can be affected by the Sun—has forced a deeper, more scientific study of the Sun, and in the process has faded the mysteries of old. Although we no longer worship the Sun in the hope that it will sustain life for another year, we study it for a similar kind of survival—the survival of our new technological standard of living. The Sun can seriously affect the performance and reliability of spaceborne and ground-based technological systems; it can endanger human life and health.

Sixty years ago the science of "space weather" did not exist. No single agency was solely committed to studying the effects of the Sun on the Earth. In probing the workings of our Sun, we've had to develop a new field of science and a new language. From "celestial" to "solar activity" to the "solar-terrestrial environment" to the "near-Earth environment" to "space weather," we struggled to define the seemingly indefinable—what exactly was the cause of the havoc on technological systems? The term "space weather" was apparently coined in a 1972 U.S. Air Force memo, though it seems to have been coined independently over the years, as several others have claimed parenthood. It leans on the analogy with meteorological weather and refers to all the space environment disturbances that result in space weather on Earth. Space weather, like terrestrial weather, has storms (radiation storms, geomagnetic storms, or radio blackouts), involves the issuance of warnings and forecasts, and pushes the development of actions that could help avoid the hazards and the disruptions of the storms. But the analogy can be misleading—this isn't rain and lightning, sunburn or flooding. Although the Sun drives all of these meteorological phenomena on Earth, the Sun also drives a more subtle space weather system: solar flares with their bursts of high-energy particles, x-rays, magnetic fields, and tremendous solar winds.

The Space Environment Center has been an important contributor to the research on space weather for more than fifty years and in 2005 celebrated its fortieth anniversary of issuing daily forecasts and warnings. It holds the distinction of being the national and world warning agency for space weather disturbances that affect people, industries, and governments around the world. Although SEC is a government agency, suggesting boring bureaucracy, the history of the organization is the history of people, science, machines, analysis, and progress. Thousands of people from all around the world have used their astute intellects to understand the science behind space weather, create the observing systems, analyze the data, and ultimately contribute to the progress of our society.

This book is, in many ways, a biography both of space weather and of the Space Environment Center. It examines the "parentage" of space weather and how our predecessors became interested in a study of the Sun (Chapter 1). The origin of SEC really occurred during World War II as the small organization attempted to bring reliable communication and navigation to soldiers and pilots (Chapter 2). In any biography scores of vital secondary characters pop up to place the subject in context. Many such characters, both international and national agencies and civilian and military organizations, contributed to and shaped space weather research and SEC (Chapter 3). SEC dramatically widened its scope thanks to the international efforts during the Space Race to develop satellites in the 1950s and 1960s (Chapter 4). A network of solar observatories around the globe played a vital role in supplying data and images that were then used to make forecasts to be sent out to the users (Chapter 5). As the agency matured, it began to develop products (not a tangible item to be sold but a range of forecasts, alerts, and packages of information) that would require constant observation of the Sun to interpret data and disseminate products (Chapter 6). The recipients of these forecasts are the organizations and industries that keep our world hooked up to power, the Internet, communications systems, and a variety of other technological necessities (Chapter 7). Throughout the life of SEC, this government agency has struggled with reorganizations and the sometimes amazing workings of the executive branch regarding funding, as can be expected in any government agency (Chapter 8).

The players in the story presented here, both from SEC and from vital partner organizations, are numerous, and they have brought this science to the forefront of space physics and solar forecasting today. Many at SEC are trying to bring space weather into the common knowledge of the next generation because they understand that Earth systems will be increasingly affected by the Sun (Chapter 9). The history of space weather is longer than one lifetime, but the science is written in the language of today and thus will be better understood by our children. This is an exciting time in the scientific understanding of our Sun and one that will affect each of us who talks on a global cell phone, finds our way with a GPS receiver, flies to the top of the world or beyond, or simply sits out under the northern lights pondering such beautiful displays.

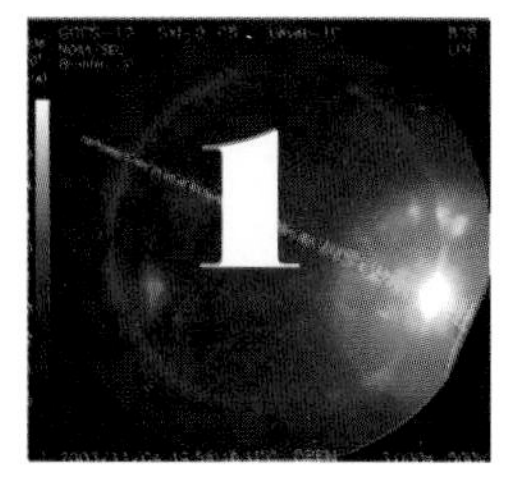

The Sun: From a God to a Star

Over the last century, archaeologists around the world have gathered on the shortest day of the year at silenced and eerie prehistoric sites to witness something few people will see—the evidence of the Sun's influence on ancient humans. At Stonehenge, in England, the midwinter Sun eases itself below the horizon at twilight, framed perfectly between the upright stones of the Great Trilithon. At Newgrange, in Ireland, the winter sunbeams strike the entrance hole in the side of the great burial mound and creep up the nineteen-meter-long tunnel to shine directly on a white stone deep inside the mound. At the observatory at Chichén Itzá, in Mexico, during sunset on the spring and autumn equinoxes, a serpent carved in stone at the bottom of the stairs throws shadows that wriggle up the stairs.

Such discoveries of ancient humans' capacity to understand astronomy exhilarate archaeologists. So it is little wonder that when they come across a slit in a rock, or a stone structure pockmarked with small openings, or a megalith set upright in the ground, they consider the possibility that the site was used in observing the Sun. Perhaps the slit marks the winter solstice, the shortest day of the year and the promise of spring. Perhaps it marks the equinoxes, when days and nights are equally long. Of course, it could be just a coincidental placing of a stone, unaligned with the Sun and of unknown purpose, but archaeologists will always search for some connection. From the earliest time, humans have known the importance of the Sun in sustaining life on Earth. In nearly every culture, we find evidence of humankind's attempt to anticipate and harness its true powers. Ancient prayers, myths and legends, wall paintings, and grand monuments give testimony to the fascination humans have had with the Sun throughout the millennia.

A short and far from exhaustive tour of the cultures through the ages and around the world provides a picture of a spiritual tie to the Sun across the cultures.

Ra, the god of the Sun, was considered by the Egyptians to be the creator of all life and was the central god in the Egyptian pantheon. Ra is represented as a man's body with a falcon's head, crowned by a solar disk (Fig. 1.1). The Egyptians believed that Ra sailed in a solar barge across the sky during the day and through the underworld at night, rising again in the east each morning. Ra was born every day and sailed a boat called *Madjet* ("becoming strong") across the twelve provinces of the sky (representing the twelve hours of daylight). When his barge dipped below the horizon, Ra died and began his voyage through the twelve hours of darkness on his night boat, called *Semektet* ("becoming weak"). On each of these journeys, Ra had to battle with his enemy Apep, a serpent that personified darkness, in order for the Sun to rise again. On stormy days, or during a solar eclipse, the Egyptians thought that Apep had triumphed over Ra and had swallowed the Sun. During the Middle Kingdom (2134–1668 B.C.), these religious beliefs developed into the state religion. The belief grew strong that humankind had been created by Ra, formed from the god's own image. The pharaoh kings linked themselves to the deity by calling themselves the "sons of Ra." At the death of the pharaoh, it was said that the monarch joined the Sun god's entourage.

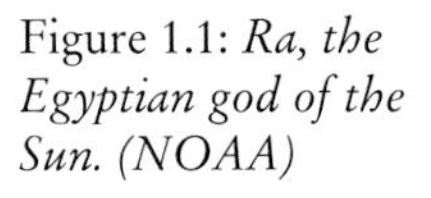

Figure 1.1: *Ra, the Egyptian god of the Sun. (NOAA)*

Archaeologists know that the Egyptians built temples to the Sun and that the pyramids were perfectly aligned with the cardinal points (north, south, east, and west) and so to some extent with the Sun. The shape of a tall, thin obelisk figured strongly in Sun worship and was often erected in the large open spaces in solar temples. Obelisks were thought to protect the temple, as the needle point at the top was said to perforate clouds and so disperse negative forces. Although these obelisks gave the Egyptians a means of showing their appreciation for the creating power of the Sun, the structures did not give them the power of predicting the Sun's movements.

The ancient Chinese worshipped the Sun, welcoming it each morning and sending it off each evening by performing rites. Ancient belief held that a three-legged crow, Sam-jok-oh, dwelt within the Sun, and the symbol of the three-legged crow ensconced in an orb represents the Sun, which is one of the twelve symbols representing imperial authority.

Oracle-bone inscriptions from the Shang Dynasty (1600–1050 B.C.) tell of Sun worship but are also the first records of solar activity, the source of space weather. The ancient Chinese were the first to observe sunspots (*ri you zhi*, literally "a blemish on the Sun"). Ancient Chinese written records sometimes referred to sunspots as a "peck measure," "small stars," or "a black cloud" seen on the Sun. Some Chinese thought that there were extra stars inhabiting the Sun, and others thought sunspots were mini–solar eclipses. Beginning around 28 B.C., they kept detailed records of sunspots, which could be seen at sunrise through the denser atmosphere (Fig. 1.2). Chinese histories contain hundreds of references to sunspot occurrences. The Chinese also observed solar prominences; in one example, such events were recorded on a tortoiseshell as "three suddenly bursting fires eating a chunk of the Sun."

Like the Egyptians, the ancient Chinese feared the evil, supernatural mischief maker that could cause solar eclipses. Some believed that it was a dragon who tried to swallow the Sun. They sacrificed to the Sun when its light dimmed, hoping that the light would return and not blink out for eternity. The possibility of the end of the world frightened the Chinese enough that they began recording solar eclipses in the hope of being able to predict them in the future. In fact, it was the ancient Chinese who made the first reliable record of a solar eclipse, in 1217 B.C.

One of the most familiar symbols of Mexico is the circular Aztec Calendar, also known as the Sun Stone. The original pre-Columbian calendar was carved from stone in A.D. 1479. The Aztec name for the

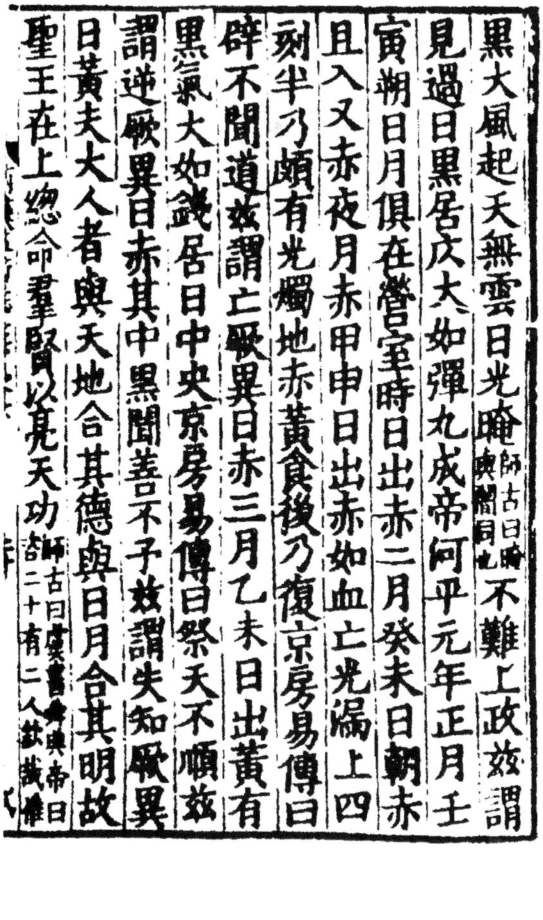

黑大風起天無雲日光晻師古曰晻與闇同也不難上政茲謂
見過日黑居仄大如彈丸成帝河平元年正月壬
寅朔日月俱在營室時日出赤二月癸未日朝赤
且入又赤夜月赤甲申日出赤如血亡光漏上四
刻半乃頗有光燭地赤黃食後乃復京房易傳曰
辟不聞道茲謂亡厥異日赤三月乙未日出黃有
黑氣大如錢居日中央京房易傳曰祭天不順茲
謂逆厥異日赤其中黑聞善不予茲謂失知厥異
日黃夫大人者與天地合其德與日月合其明故
聖王在上總命羣賢以亮天功師古曰虞書舜典帝曰咨二十有二人欽哉惟

Figure 1.2: *Written Chinese record of a sunspot event from 28* B.C.*, in Han Shu, Wuxing zhi.*

stone is Cuauhxicalli, meaning "eagle bowl." It was dedicated to Tonatiuah, the Aztec deity of the Sun, and showed the Aztecs' belief in cyclical time. Like the Egyptians, the Aztecs believed the Sun to be the creator of all life. The Sun Stone calendar weighs an impressive twenty-five tons, has a diameter of almost twelve feet, is three feet thick, and is beautifully adorned with glyphs, geometric figures, and grinning faces (Fig. 1.3).

Figure 1.3: *The Aztecs used the Sun Stone, this famous Aztec Calendar, to carefully track the solar year. The calendar also symbolized the Aztec worship of the creator of all life, the Sun. (NOAA)*

There are two parts to the calendar, each one having a particular function. The first part is called the *tonalpohualli* ("counting of days"), a cycle of 260 days. Diviners used this part of the calendar to predict whether they would have victory in war or a successful season for the crops. The second part of the system is the *xiuhpohualli* ("counting of years"), a 365-day calendar based on solar days that counted up to a year. Like synchronized gears, the two calendar cycles work together, forming what the analogy to a century would be: fifty-two years. Aztecs celebrated the alignment of these two calendars with festivals and religious ceremonies, including the legendary human sacrifices.

The Polish cleric Nicolaus Copernicus, 1473–1543, lived a near-cloistered academic life in which he developed an intense interest in astronomy. He made careful observations of the Sun (without any optical aids) and concluded that Earth rotated on its axis once daily and traveled around the Sun once yearly—a fantastic concept for the time. Copernicus's heliocentric theory countered the belief that had been firmly held by scientists of the Western world since A.D. 150. Claudius Ptolemy, an Egyptian living in Alexandria, had asserted that the universe was a closed sphere that enveloped all planets and stars. The stars were fixed to the inside "wall" of the sphere, and all the planets moved in fixed relation to each other. At the center of the sphere was God's Earth, and everything revolved around it. Beyond this sphere was nothing. Ptolemy himself built upon the ideas of Aristotle (384–322 B.C.), who also believed in an Earth-centered universe. Aristotle's universe had another distinguishing characteristic—it was perfect. Planets moved in perfectly circular orbits, and celestial bodies were flawless spheres. In the face of a thousand-year "truth," Copernicus's theories were bound to spark a raging controversy fifty years later.

Copernicus's work was published in 1543, just before his death. His initial writings and descriptions of his theories had been circulated earlier, but few people read or understood them. Not only did Copernicus's work suggest a heliocentric model of the known universe, but one of his students, Theophrastis, observed and recorded sunspots, giving credence to the theory that the Sun was not "perfect." Both theory and observation suggested that humankind was a part of the imperfect universe, not dominating at the center and the "next thing to God." These ideas set the stage for scientists and theologians to clash in the following century.

By the late sixteenth century, the Roman Catholic Church, already passionately opposed to any belief that countered Aristotle's Earth-centered (thus human-centered) teachings, faced a new threat in the form of two outspoken Italian scientists. Giordano Bruno and Galileo Galilei fully embraced Copernicus's Sun-centered views of the solar system. Bruno dared to suggest that space was boundless and that the Sun and planets were but a single system among many in the universe. The door was now open for speculations that there might be other sentient beings like, or superior to, humans. This was too much for the Roman Catholic Church. Bruno was tried by the Inquisition and burned at the stake in 1600.

Galileo, who agreed with many of Bruno's theses, developed a powerful telescope with which he observed the heavens. In his keen observations of the solar disk, he noted the appearance of "blemishes" (or sunspots), which, again, ran counter to Aristotle's teachings that celestial bodies were immutable. Galileo was outspoken and staunch in his beliefs and thus faced a trial similar to Bruno's thirty-three years earlier. Galileo's forced renunciation of Copernican theories kept him alive, and he was allowed to remain imprisoned in his house until his death in 1642. Despite his struggles with the church, Galileo enjoyed the support of several very powerful clergymen, including one admiring pope. Galileo was highly regarded by church and lay academics, as well as aristocrats, for being a prolific and formidable astronomer and physicist. His reputation likely influenced the church in sparing his life.

Galileo's early observations of the Sun were invaluable to the eventual development of solar physics. Most of his early observations were done with the naked eye and small telescopes. At first glance, the Sun looked flat, and there were only guesses as to what the surface was like. But as soon as Galileo began to scrutinize the Sun, he quickly saw that it had sunspots on its surface, and that it was a sphere that rotated, because the spots slowly moved westward across the surface and disappeared only to reappear fifteen days later on the east side. He made detailed drawings of sunspots on consecutive days, which showed both the evolution of sunspots and the Sun's rotation rate. Galileo's records from his studies still exist but are not extensive (Fig. 1.4). Careful observations from 1600–1626, done by the German astronomer Friar Scheiner, were the first continuous solar observations of scientific value. Scheiner began the consistent observation of sunspots that has become a foundational historical record in the field of space weather.

As a final sample of different cultures' interest in the Sun, the astronomical observatory at Jaipur, India, is notable in that it embraced both Eastern and Western cultures. Around the year 1720, the ruling maharajah, Sawai Jai Singh II, built the unique observatory. He ascended to the throne of amber at a young age but later abdicated to devote his life entirely to his love of the sciences and the arts. The Jantar Mantar site at Jaipur, some 200 km (125 miles) south of New Delhi, is one of the great accomplishments of modern astronomy. The building style is traditional Hindu, with some Greek and Islamic influences. Around the main observatory building lie enormous astronomical instruments built of stone—equatorial and hemispherical sundials, instruments to measure altitude and azimuth, an astrolabe, and many more (Fig. 1.5). Those unaware of the highly scientific function of these instruments might be inclined to think a giant had dropped his toy building blocks: staircases leading to nowhere, arches made of massive stone blocks, concave ramps littering the field in front of the main building. These unique instruments and the use of European and Islamic scientific method at Jaipur make the observatory a keystone of astrological and astronomical study. Using the monumental stone instruments, one can track the motion of the Sun or stars to the second.

As far back as we can document—and we can suppose further back—people have worshipped and watched a phenomenon seemingly unrelated to the Sun—the beautiful northern lights, or aurora. These

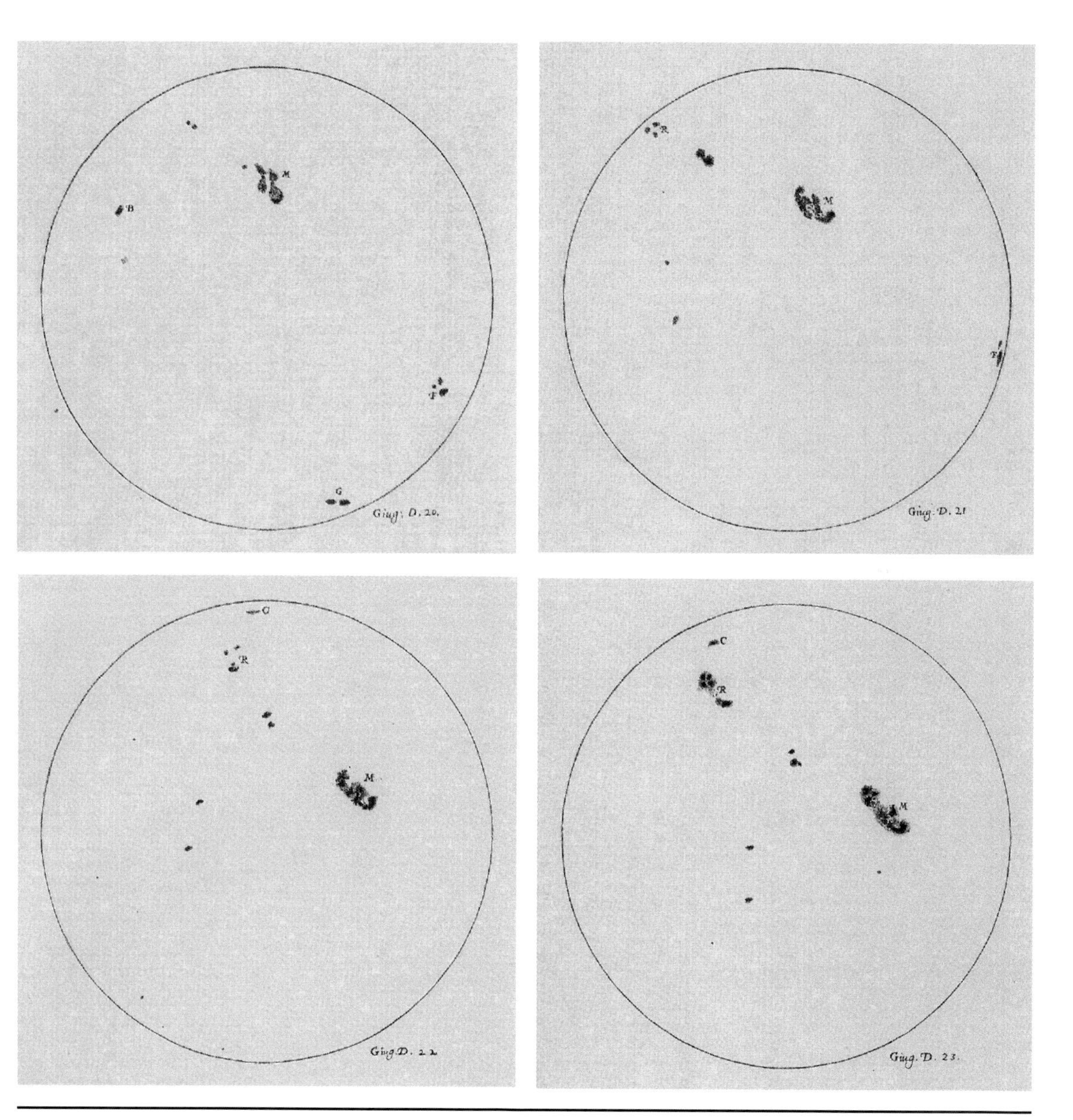

Figure 1.4: This *consecutive series of sunspot observations was made by Galileo in the summer of 1613, and published in* Istoria e Dimostrazioni Intorno Alle Macchie Solari e Loro Accidenti Rome *(History and Demonstrations Concerning Sunspots and Their Properties). In these drawings Galileo has documented the rotation of the Sun, showing the sunspots traversing the surface. Note that the Sun's rotation axis is not vertical in these drawings. (Original publication scanned by Rice University)*

stunning, eerie, and silent displays of color in the night sky have captivated people from well before recorded history. Auroras sweep across the sky like waving curtains or curling swirls, lasting sometimes for hours, then disappearing; they were beyond the comprehension of early observers. Since auroras are most likely seen in the polar regions, where they are made even more impressive by the long hours of dark-

Figure 1.5: *Jantar Mantar, the large observatory complex at Jaipur, contains numerous sundials and astrology devices. The largest sundial in the world, accurate to within a fraction of a second, is a main feature of the observatory. For scale, you can just make out a person in the top right-hand corner. (Michael Grant and Penelope Lewins, www.grant.org)*

ness in a winter day, people living in the higher latitudes were driven to explain the origin of the lights. For thousands of years or more, Eskimos thought the aurora showed them a view of heaven, where fiery, beautiful light bathed newcomers in a warm welcome to the afterlife. The Lapps saw the northern lights as powerful and fierce and respected and feared them as messengers of God. Scorched reindeer-skin jackets occasionally found by Lapps were thought to be all that was left of a man who had provoked the messengers. Although rarely seen at lower latitudes, the awesome light displays sometimes appeared in the European night sky, above Paris or Rome. The lights seen at lower latitudes were "blood" red and during the Middle Ages were believed to portend war, plague, or fire.

The mystery of the aurora would remain veiled until Kristian Birkeland, a Norwegian scientist interested in magnetism, made his definitive study of the aurora in 1900–1901 (Fig. 1.6). Birkeland's predecessors believed that either magnetic, electric, or cosmic forces caused the aurora, but few satisfactory theories were developed. Ultimately, Birkeland would develop a coherent theory that combined all of these forces and explained the phenomenon.

Birkeland had always been fascinated with magnetism and the work done by the Englishman William Gilbert in 1600. Gilbert's three main discoveries on magnetism were revolutionary: first, the Earth was a giant magnet; second, Earth had magnetic north and south poles (not to be confused with the geographic north and south poles); finally, magnetic field lines spread out from the poles in the shape of a halved apple. The magnetic field surrounding the Earth would come to be called the "magnetosphere." This information on the first of the three possible causal forces—magnetic, electric, and cosmic—would prove infinitely valuable to Birkeland as he studied the aurora. The second force suspected to influence the aurora, electricity, was linked to magnetism by the works of Hans Ørsted (1819), Michael Faraday (1843), and James Maxwell (1855).

The final force, cosmic influence, Birkeland would study on his own. Armed with the scientific findings of Gilbert, Ørsted, Faraday, and Maxwell and living so far north, Birkeland witnessed the aurora and measured the strength of the magnetic fields near the magnetic north pole using a magnetometer. The magnetometer could measure fluctuations in the Earth's magnetic field and made it possible to infer the presence of a vast electric current system driven by the magnetic field lines in the polar regions. Birkeland wrote, "The currents pass, on average, at a height of approximately 100 km [62 miles] above the terrestrial surface and can cause strong magnetic perturbations of a total intensity above 400,000 amperes." But Birkeland knew that just describing the magnetic field and the electric current was not enough to explain the

Figure 1.6: *Kristian Birkeland, a Norwegian inventor and scientist, seated in front of his assistants and a Samoyed guide before their second study of the aurora from Novaya Zemlya. (Norwegian Technical Museum, Oslo)*

northern lights. To find the true cause, he would have to find the source of the light. He turned to the possibility of cosmic influences: possibly the Sun. But how? If charged particles from the Sun did stir up the electric currents and cause the aurora, it seemed impossible that they could travel the vast distance between Sun and Earth. Much more research was needed.

Birkeland and his small team of students carried out a dangerous research mission, climbing to the tiny auroral observatory on Haldde Mountain in Norway during the winter. The hurricane-force winds, terrible snows, frigid temperatures, and near-total darkness were trying on the men. Trips to and from the mountain required dealing with semiwild reindeer pulling sleds, avalanches and deep snow, and frostbite and resulted in serious injury or death for some. But what the scientists suffered in extreme living conditions, they compensated for with pleasant companionship—and the definitive answer of the aurora.

With knowledge gained on this research trip, Birkeland began to understand the cosmic influences on the aurora. He began to suspect the Sun after he made several observations of sunspots or solar flares from three days to eighteen hours in advance of aurora sightings. Others, in particular the revered Lord Kelvin, did not accept the idea that charged particles from the Sun could travel so great a distance. Birkeland could not explain how the particles traveled the enormous distance, but he knew they did not originate on Earth, and they had to come from somewhere. The correlation between sunspots and solar flares and the aurora was the best proof he could furnish. Despite the stern academic doubts about his scientific theories, Birkeland published the first realistic theory of the northern lights. He argued that electrically charged particles ejected from the Sun are captured by Earth's magnetic field and directed toward the polar regions. Once in the magnetic field, the particles collide with atoms in the atmosphere, exciting the atoms, which absorb electrons and consequently emit different colors of light (Fig. 1.7). He was largely correct and one of the first to describe a geomagnetic storm with understanding.

One tool that Birkeland (and Gilbert before him) found extremely useful in examining the magnetic structure of the Earth was the Terella. As the name suggests, the Terella was a simple model of the Earth complete with a magnetic core. In the model, an orb with an iron core wrapped with copper wire and a phosphorescent outer layer, mounted slightly off the vertical, mimicked the Earth. The orb was placed in a vacuum chamber and bombarded with electrons. When the electrons entered the chamber, they spiraled from their source toward the magnetic poles of the orb, ionizing the air in the chamber, and created glowing halos around the poles, just like the aurora. These experiments perfectly repro-

Figure 1.7: *The northern lights are more formally called the aurora borealis (which is somewhat of a misnomer because* aurora *means "dawn," and the aurora are unrelated to sunrise); the southern lights, aurora australis (so named by Captain Cook in his 1770 exploration of the oceans around the Antarctic), occur at the same time as the northern ones, linked by the same geomagnetic phenomenon. (Dave Parkhurst, ©2004, www.TheAlaskaCollection.com)*

duced what Birkeland suspected happened in the Earth's atmosphere to cause the aurora. In 1930, Sydney Chapman and Vincent Ferraro in England further verified and refined Birkeland's theory, crediting plasma clouds expelled from the Sun and interacting with Earth's magnetic field as the mechanism for triggering polar auroras. Thanks to the research of all of these men, we now know that auroras are one of the few visible manifestations of space weather.

Incidentally, Birkeland's life lends quite a bit of color to the field of astronomical research. He was a fearless and perhaps foolhardy experimentalist and inventor. As a student, he lost part of his hearing during an experiment with radio noise, but this disability did not seem to faze him. Later he attempted to build a rail gun that shot a missile using electricity along a conducting rail lined with magnets. His *kanon* short-circuited and exploded, producing nitric oxide as a by-product of this intense release of energy. Instead of being humbled by the explosion, he found his interest piqued by the nitric acid such an explosion could create, as he was keen on inventing a feasible way of creating fertilizer. At this time, around 1904, natural fertilizer was in short supply, and furnaces that could produce enough energy to mass-produce the main ingredient—nitric oxide—were just being developed. Birkeland had high hopes of making his fortune in fertilizer, but they were dashed by wily investors. Birkeland was certainly a passionate scientist and a likable man, though he was troubled by bouts of mental illness.

Birkeland's early research centered on the creation and nature of the aurora, but it did not shed light upon another question—if and how auroras and weather were linked. For centuries, people living in northern latitudes desired expert knowledge about the weather for fishing and agriculture. Many different cultures—people of Scandinavia and Greenland, the Cheyenne Indians in Wyoming, the Penobscot Indians in Maine—believed they could use the northern lights as predictors of the weather. Although Europeans, centuries ago, thought the aurora presaged calamitous human events, others who saw it more frequently thought the lights could be trusted for day-to-day predictions. Did the aurora bring on or otherwise influence terrestrial weather, causing notorious storms, pounding rains, or lightning strikes? Observers sometimes saw a storm not long after large auroral displays, suggesting a correlation between the two. Research into the possible link between the aurora and weather would be taken up in earnest as modern scientists gained a more complete understanding of Sun-Earth interactions.

As in most fields of modern science, developments in the past sixty years have dramatically increased what we know about Sun-Earth interaction, including the aurora. Before that time, our observations of events on Earth had become fairly sophisticated, but scientists had a long way to go to observe and understand the Sun. The sophistication and depth of our current knowledge about the Sun remove the star from its occult sphere of worship and mystery and give us a modern, scientific understanding our ancestors could not have conceived of.

Our Sun is a medium-sized star, four billion years old, that stays a comfortable ninety-three million miles away as we slowly circle around it in 365 days. Earth, the third planet from the Sun, relies on the Sun for everything: Earth's orbit and rotation, weather, geology, life, day and night. Most people know certain key facts about the Earth: its daily rotation, its yearly precession around the Sun, its tilted axis that creates summer and winter, and its magnetic poles. However, many are surprised to learn that the Sun has similar features. The Sun also rotates around its axis. An Earth day, measured by looking at the rising and setting of the Sun, is approximately twenty-four hours. A Sun "day" is approximately twenty-seven Earth days, as determined by pinpointing a position on the Sun and measuring how long it takes for the same point to appear after the Sun makes one full rotation. The Sun has a cycle similar to Earth's year. Scientists identify the cycle of solar activity (analogous to a season, one might say) by counting sunspots (Fig. 1.8). The Sun's "yearly" cycle of solar activity changes from Solar Minimum, with the fewest number of sunspots, to Solar Maximum, with the greatest number of sunspots, and back to Solar Minimum again (Fig. 1.9). A solar cycle, from Minimum to Minimum, is roughly eleven Earth years. Scientists number the eleven-year cycles (we are in Solar Cycle 23 in 2005), and there is also a secondary cycle. The Sun completes a reversal of its magnetic poles from one solar cycle to another, so a complete change back to the start of the magnetic cycle can be described as a twenty-two-year cycle. During the solar cycle, varying solar weather is brewed by the Sun, much like the seasons here on Earth, so it would be misleading to think that solar weather is near nonexistent during Solar Minimum. Rather, the solar weather changes in character.

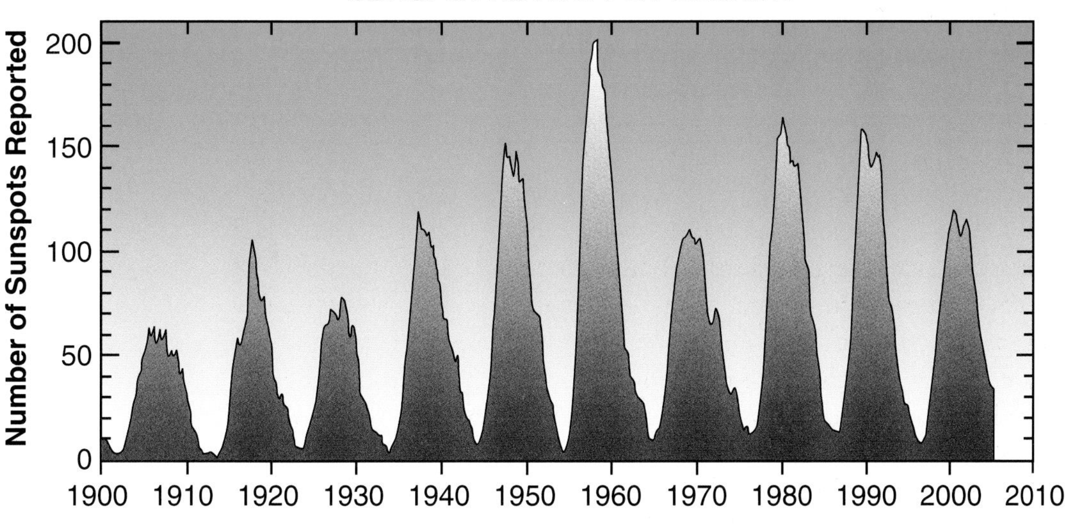

Figure 1.8: *The sunspot cycle for the last ten cycles is plotted, in this case, by monthly number of sunspots. For centuries the sunspot cycle has been, and continues to be, the standard measure for solar activity, even though some types of solar activity are offset or anticorrelated with sunspot numbers. (NOAA)*

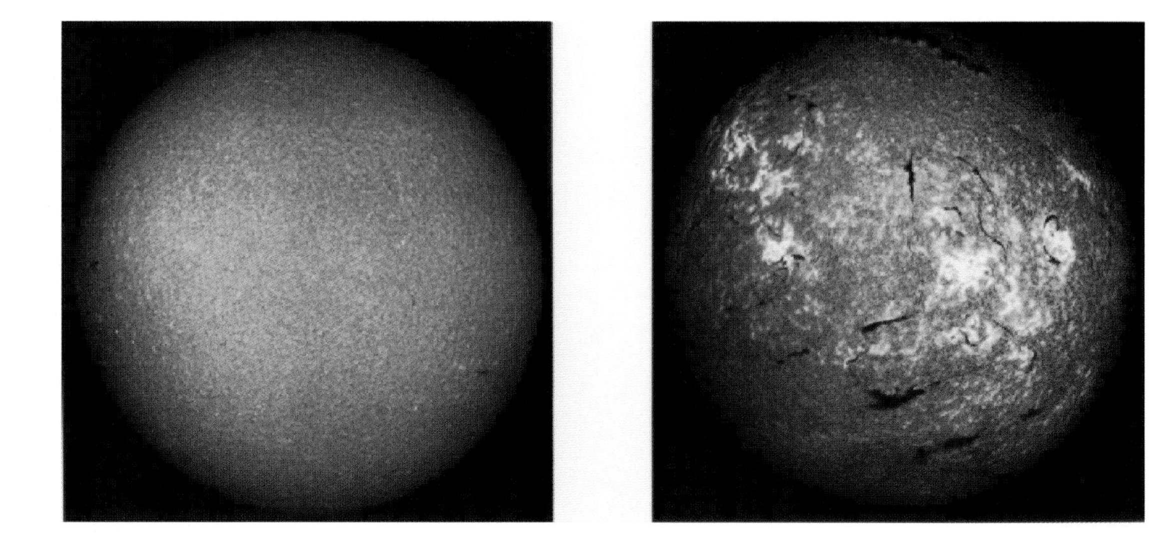

Figure 1.9: *The Sun at Solar Minimum and Solar Maximum looks quite different in H-Alpha images. Sunspots are seen during Solar Minimum, but with much less frequency. (Big Bear Solar Observatory, California Institute of Technology [now operated by New Jersey Institute of Technology])*

People think about the Sun as they see it, typically as visible light. Some of the Sun's energy is in the visible range of the electromagnetic spectrum. This energy influences our weather and helps sustain life on Earth, but the other parts of the spectrum, not in the visible range, are of greater interest to space physicists. Wavelengths outside the range of visible light can tell scientists a great deal about the structure of the Sun. Like that of the Earth, the Sun's atmosphere is composed of several layers, each having a different density and temperature: the corona is the uppermost (outermost) layer, followed by the chromosphere and photosphere (Fig. 1.10). Each layer emits different wavelengths of light. Using instruments and filters that admit only a certain wavelength, scientists can view the Sun's various layers. An instrument that measures x-rays would "see" the corona, while an ultraviolet instrument would take data from the chromosphere, and wavelengths in the visible-light range would reveal the photosphere. These filters have allowed scientists to know a great deal about solar activity at every level of the solar surface.

Ultimately, the greatest benefit that comes from knowing so much about the Sun is an understanding of how the Sun influences the Earth. The Sun and the Earth have a somewhat paradoxical relationship—life would not exist on Earth without the Sun, and yet the Sun's radiation is deadly. Fortunately, the Earth has a mighty and complex shield that serves as protection from the Sun—a magnetic field. This magnetosphere stretches around the Earth, running from the south to the north magnetic pole, somewhat like longitude lines on a globe. Unlike longitude lines, the magnetic field lines do not run just over the surface of the Earth but stretch up well beyond the atmosphere. The solar wind—the outward flow of magnetic field and plasma (photons, ions, and

Figure 1.10: *A cross-section of the Sun showing the structure of the interior of the Sun (right), as well as the surface features present in the photosphere (left). (NOAA)*

subatomic particles) from the Sun—distorts the Earth's magnetic field lines (Fig. 1.11). As Birkeland demonstrated, it is the solar wind interacting with the magnetosphere that causes the aurora.

Volumes have been written about our knowledge of the Sun and the complexity of Sun-Earth interaction. But the purpose of this book is to examine only the Sun-Earth interactions that directly affect human beings. The Sun's effect on human systems has gradually come to be known as "space weather."

Except for the beauty of the aurora, most people were untouched by the great storms of particles and energy ejected from the Sun and carried to Earth by the solar wind. As early as the 1600s, explorers experienced occasional trouble with their compasses, which would follow the distorted magnetic field lines and not point to the magnetic north during a geomagnetic storm, but the impact was not far-reaching. The first signs of space weather having an impact on the general public began in the mid-1800s with the advent of the telegraph. The long conducting wires that stretched across the country were large enough to be impacted by solar storms. In 1859, highly charged particles entered the polar regions, as happens during auroral sightings, and caused a geomagnetic storm (disruption of the Earth's magnetic field due to charged particles and waves of magnetic field from the Sun, carried on the solar wind). Telegraph wires required electricity or a battery to carry the signal from one station to the next. Disturbances in the atmosphere were

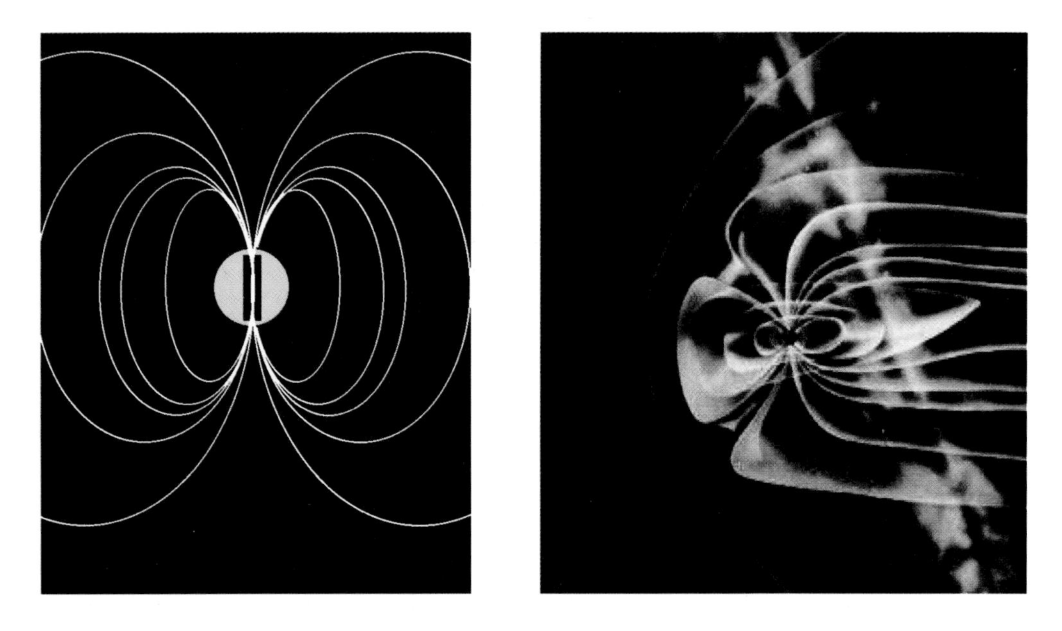

Figure 1.11: *(Left) Without disturbance, the Earth's magnetic field would look like the field produced by an ordinary bar magnet. (Right) The solar wind distorts Earth's magnetic field, blowing the field into a swept-back tail. (NOAA)*

often large enough to cause a current to flow in the telegraph wires with enough voltage to allow operators to transmit signals even after disconnecting the wires from the batteries. This puzzling and surprising phenomenon was dubbed the "celestial battery," implying that it was an otherworldly phenomenon. The name was apt because the phenomenon was caused by a celestial body (though no one knew it was the Sun) and brought the power of a battery to the wires without artificial generation. Although noted by telegraph operators, this phenomenon had little effect on telegraph users because it happened infrequently, and disruptions in the telegraph service were, unfortunately, not unusual.

New technologies revealed new problems. In 1902, Guglielmo Marconi successfully sent a radio signal across the Atlantic Ocean, though he was not exactly sure how it worked. Oliver Heaviside, a telegrapher and physicist, correctly theorized that radio waves reflect off an electrically conducting layer at the top of the atmosphere, essentially forming a two-line transmission path: one up to what R. Watson-Watt called "the ionosphere" and one down to the receiver on the ground. But like the magnetosphere, the ionosphere is sensitive to solar emissions, rendering radio communication vulnerable to space weather.

In the early years of radio technology, few people besides the pioneers used or knew about high-frequency radio. A remarkable architectural feat would soon aid in opening radio broadcasting to a larger world. The Eiffel Tower had been built for the International Exhibition of Paris of 1889 to commemorate the centenary of the French Revolution (Fig. 1.12). In 1909, having served as an exhibition piece for twenty years, the tower was threatened with demolition. Luckily, it was saved because of its antenna, which was used for telegraphy at that time. Soon-to-be news service broadcasters successfully argued that the tower served a vital role as a broadcast antenna for telegraphy and radio transmissions. Marconi, later dubbed the "Father of Radio," had capitalized on the tower's great height for transmission of radio waves when he invented what he called the "wireless telegraph."

With increased radio use, people started noticing disturbances in their radio signals. In the 1920s, scientific studies on radio-wave propagation indicated that solar interference with the ionosphere, possibly originating from sunspots, could cause occasional annoying hissing noises, "like bacon frying." At other times, the radio signal would fade and the broadcast become inaudible. Marconi had come to realize the tenuous quality of communications in the event of solar disturbances. Upon the tall, impressive Eiffel Tower he erected antennas that could monitor space weather–caused radio interruptions. The tower was thus needed to broadcast signals and also to predict when the signals might be disrupted. The first space weather forecasts were issued from the Eiffel Tower in 1933, the beginning of crude space weather forecasting. In a way, the Eiffel Tower stands today thanks to its role in space weather forecasting, and space weather owes a debt to the tower for its role in launching this field.

Figure 1.12: *The Eiffel Tower, built for the International Exhibition in 1889, proved useful in transmitting long-distance radio signals. (Jeffery Howe, Boston College)*

The ionosphere would be the main focus of space weather concerns for the next half century, and during that time scientists were learning about this invisible region. Roughly the third atmospheric layer from the ground, the ionosphere is considered part of the thermosphere. This layer is filled with charged particles that originate when cosmic x-rays ionize the neutral atmosphere. The ionosphere, with its varying density of particles, interacts with high-frequency radio waves in three different ways: radio waves can pass straight through, scatter in all directions, or reflect back to

Earth (Fig. 1.13). This final case allows for radio transmission. In a stable atmosphere, broadcasters can target locations in the ionosphere that will reflect a signal to a desired receiver. However, the particles in the ionosphere are sensitive to emissions from the Sun; for example, during a solar radiation storm, increased radiation alters the net charge in the ionosphere. As a result, the reliable, reflecting parts of the ionosphere are disturbed, leaving the radio waves to either pass through, scatter randomly, or bounce off at unpredictable angles. In any of these cases, the radio signals never arrive at the expected receiver location. In 1926, the distance to this conducting layer was determined by measuring the time it took for a radio signal to be sent straight up and bounce back.

The use of radio and the Eiffel Tower were just a beginning, the catalysts for greater understanding of how the Sun affected people. The age of space weather—the in-depth study of the Earth's response to solar phenomena, not just the study of the Sun or the observation of the aurora—was about to begin. As for so many things, the driver for this new research was to be war. World War II reached around the globe and brought together countries that had a common need to improve communication, navigation, and weather forecasting for the sake of their thousands of soldiers in the field. War made the need for knowledge about the Sun global and urgent.

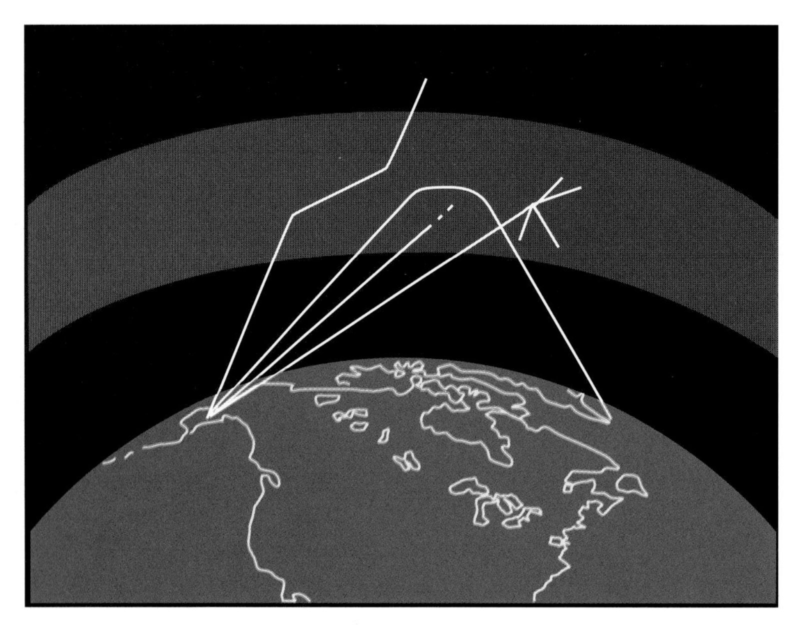

Figure 1.13: *Radio transmission can behave in several different ways when it hits the ionosphere: signals can penetrate the ionosphere and refract, or can be absorbed and be lost, or can hit dense patches and be scattered. Signals can bounce in predictable ways when the ionosphere is quiet but move less predictably during space weather storms. (NOAA)*

War!

Lucky soldiers have much to be grateful for when they finally return home, alive, to the arms of their families. They managed to dodge the bullets, there was enough food and sufficient shelter, the generals made the right decisions, or the enemy failed to see them coming. Few perhaps realize that intelligence about solar activity is among these life-saving factors. The Sun does not affect mortars or musket balls, and as recently as the early twentieth century, solar activity was hardly noticed by military leaders except for the aurora, telegraph signal interruptions, and disruption of still-novel radio communication. It would take developments in technology and a threat to human life before space weather research would become an important part of twentieth-century warfare. World War II had both of these necessary factors. Unbeknownst to most soldiers, the rapid developments in solar research during this period would play a part in bringing them home to their loved ones.

War is critically dependent on weather. Weather often proves disastrous for troops in the field—ships sink in violent storms at sea, armies hunker down against bitter winters, too much rain stops vehicles and threatens to rot feet, too little water brings the possibility of fatal dehydration. History books tell of the enormous part played by weather in the outcomes of battles, if not wars. More subtle and less visible, space weather played an important part in World War II because of its relationship to communication, navigation, and, indirectly, terrestrial weather.

U.S. military officials during World War II were concerned about solar activity disrupting vital radio communication and navigation systems. The forty years between Marconi's development of radio and widespread use of the technology during World War II saw a leap in knowledge about radio technology and the ionosphere. In order to understand the problems with communication during the war, it is useful to know where radio technology stood as the conflict opened.

Early radio communications used line-of-sight transmission. In other words, radio signals could be passed only between two receiving towers without obstacles between them. At greater distances, straight line-of-sight signals would not follow the Earth's curvature but would follow a line at a tangent to the surface and be lost. When much longer-distance communications became important from, say, the United States to Europe, line of sight was of no use. In 1925 Gregory Breit and Merle Tuve in the United States and Edward Appleton and Miles Barnett in Great Britain showed for the first time that radio waves could be reliably reflected from the ionized portion of the atmosphere (the ionosphere). After this discovery, high-frequency (HF) radio would offer valuable long-distance communications. Such communications could be sent over vast distances, even around the world under certain conditions.

In its usual state, the ionosphere is predictable, but under disturbed conditions, HF communications can be problematic. HF radio operators knew how to tune the frequencies to hear from, and send messages to, known places, but radio operators were also aware of ionospheric disturbance, as their signals would sometimes "fade," or become inaudible. Often during ionospheric disturbances, the radio signal would bounce at an unexpected angle and end up in a different location than expected. In peacetime, this was a nuisance. In wartime, not knowing where a signal would end up was risky.

Scientists knew that above a certain maximum frequency, HF radio signals would pass right through the ionosphere into space and be lost forever. A signal sent at too low a frequency would yield only static. The higher the frequency, the better the transmission, but only up to the maximum. In the late 1930s, Newbern Smith (later the head of the Interservice Radio Propagation Laboratory [IRPL]) and his team developed a model for determining maximum usable frequencies (MUFs). This model of the best MUF in a "normal" ionosphere could be adjusted based on space weather disturbance measurements. These models pinpointed the perfect midrange frequency that was neither too low nor too high (Fig. 2.1).

In order to know the MUF, Smith had to know the state of the ionosphere. He measured the location of the bottom of the ionosphere by a vertical ionogram: sending a radio signal straight up and seeing how long it took to return after bouncing off the ionosphere. Oblique ionograms were similar to vertical ionograms, except that the signal was sent at an angle and received by a different receiver. With these data, Smith calculated the height of the ionosphere and, by changing the frequency, was able to calculate the MUF.

Of course, the state of the ionosphere is not constant, especially during solar activity. Scientists could observe solar activity and change the MUF predictions based on what they knew about ionospheric disturbances. Since the MUF was not constant from day to day, radio operators could not just tune their radios to the correct frequency and hope to communicate with their desired party. For any kind of reliable transmission, operators had to listen to predictions of the MUF and agree with the

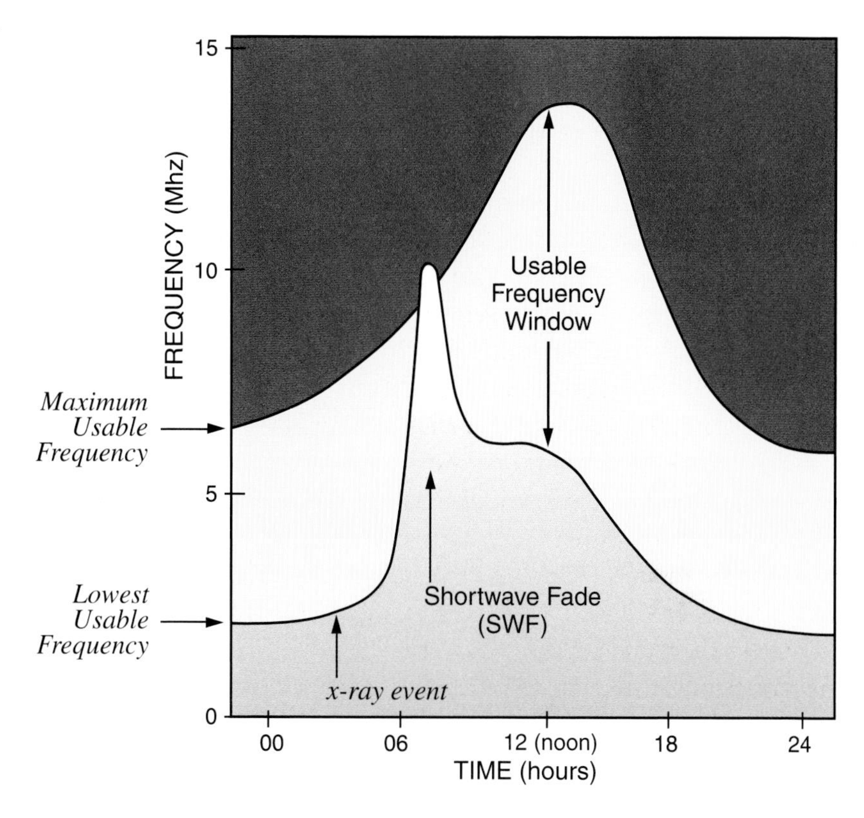

Figure 2.1: *The maximum usable frequency (MUF) is the highest frequency that will allow HF radio communications to advantageously use the ionosphere. The lowest usable frequency (LUF) is the limit below which the radio signal will be too weak to travel long distances. Space weather can affect the MUF and LUF so that when the MUF is low and the LUF is high, the available wavelength for favorable transmission shrinks to a point at which signals are completely blocked, causing a shortwave fade. (NOAA)*

party with whom they wished to communicate that, say, next Tuesday they would broadcast at a particular frequency. The MUF would become one of the most important parameters for HF transmission during World War II because it was vital that radio signals traveled to the right receivers and were audible.

By the opening of World War II, radio communications were essential between command centers and field troops and among international allies. Such a global war required communications among battleships, submarines, air squadrons, foot-soldier commands, and national and international supreme commands. Often plans could be made only after headquarters had been apprised of a situation from the field and the field had received the overall strategy from headquarters. Radio messages had to travel reliably from sender to receiver, and the signal had to be understandable (not all static). These communications meant life and death to the soldiers. However, both Allied and Axis powers had to contend with space weather geomagnetic storms that could disturb radio. Poor communication quality could lead to radio signals being picked up by the enemy, orders being lost, vital information from the field not reaching headquarters, or, among other possibilities, commanders not getting terrestrial weather reports.

During World War II, everyone relied heavily on weather predictions. Major military strategy hinged on the weather; Britain even classified weather information as Top Secret. Unfortunately, such secrecy made any transmission of weather forecasts to distant Allied weathermen nearly impossible. Weathermen in Greenland, say, could make only local forecasts based on limited data. One enterprising British team took to reading the sports pages from a U.S. newspaper. If a New York baseball game had been rained out, they could infer that a low-pressure system was heading their way across the Atlantic.

Although limiting, classification of weather forecasts was understandable. The greatest offensive in World War II depended on the weather: D-day was originally scheduled for June 5, 1945, but conditions were terrible for a water landing. High winds caused choppy waters in the English Channel and brought in fog that would hinder the air bombing scheduled to take place before the invasion. General Eisenhower was reluctant to believe the weather forecasters, who said the weather would break on June 6. A hasty move could cost soldiers' lives, but delay could lose the value of surprise and could also lead to unnecessary loss of life. He reluctantly ordered the water landing to take place on June 6, and to his surprise and relief, the weather improved as predicted (Fig. 2.2). There can hardly be any doubt that terrible weather would have caused countless sick and drowned soldiers as well as the possible failure of the invasion. Good forecasts and good dissemination of the information were key. Space weather could hinder that process by disturbing radio signals.

The discovery of the sunken German U-boat, *U-869*, off the shores of New Jersey in 1991 illustrates space weather's negative effects on radio communication. According to World War II records from Germany and the U.S. archives, the submarine had orders to proceed to the New York area from the German shipyards of Bremen in December 1944. The commander chose to travel far to the north and then down the Canadian coast rather than risk the open, well-patrolled Atlantic Ocean. During the voyage, officials at the German control center waited for the first routine check-in from the sub and began to worry when they heard nothing. They decided to reroute the sub to Gibraltar. Radio communication was dangerous for a sub, which had to surface to send a message and could be picked up by enemy radar while at the surface. This may explain why the commander chose not to surface and check in. The submarine never diverted its course to Gibraltar but headed on to New York. Some speculate that the commander was disobeying orders, but it is more likely that the message containing the change of orders was never received. A plausible reason why the message may have failed to get through was that space weather

Figure 2.2: *The D-day invasion depended on both surprise and good (enough) weather. Originally the invasion was scheduled to take place on June 5, 1944, but poor conditions delayed it. Weather forecasters predicted that the next day would provide better weather, so General Eisenhower, with some reluctance, rescheduled the invasion for June 6. The forecast proved accurate, and the invasion was successful. (www.archives.gov)*

activity at the time was high enough to have disturbed the ionosphere. If the radio transmission indeed failed, the missed communication led to the loss of the German submarine and all hands in American waters because U.S. forces had intercepted the U-boat messages and were able to intercept the sub.

Like radio communication, navigational tools were a necessity, especially for pilots flying over featureless terrain such as the Atlantic Ocean. The loss of reliable navigational tools often meant the loss of the plane and crew. Relatively new radar technology was susceptible to space weather interference. Indeed, the age-old compass technology was prone to misreadings too. The compass needle, usually pointing toward magnetic north, will respond to the north-pointing magnetic field line that is distorted away from north by a geomagnetic disturbance. Although there is no documented proof, this same compass malfunction no doubt occurred throughout the age of ocean exploration. Navigational errors were fairly common, and sailors may or may not have been aware that problems could be due to a misleading compass. Pilots during World War II possessed virtually the same compass technology as early explorers, but the consequences for navigational errors were even more serious. Getting slightly off course by following a malfunctioning compass could mean an accidental, potentially fatal trip into hostile territory.

Radar was a specialized application of radio signals, and as such became unreliable during ionospheric disturbances. The first radar system was designed by two British scientists, Robert Watson-Watt and Arnold Wilkins. They successfully constructed a system that bounced radio signals off an airborne plane to reveal the plane's location. The first of six or seven radar stations was in operation by August 1936. By the start of the war, both the British and the Germans used radar systems, but it wasn't until 1942 that the systems became sophisticated enough to be really useful. Substantial developments improved the range so that the British, for example, could watch for German bombers flying over the English Channel. But like all radio signals, the accuracy of radar was jeopardized by geomagnetic storms. During geomagnetic disturbances, operators sometimes "saw" German planes where none existed.

In the 1940s, with the vulnerability of aircraft radar and compass systems, accidents were inevitable. In February 1944, after an attack by British bombers on Germany, British spies in Germany reported how many planes had been shot down, and therefore how many had left safely to return to England. However, one plane did not return. There had been no radio communication with the crew, and no one had seen the plane go down. No wreckage was ever recovered. This was a time of high geomagnetic storms, and the missing crew was assumed lost to the North Sea or the Atlantic Ocean, presumably because of navigation equipment failure and inability to use the radio.

As a result of ionospheric problems that interfered with radio and navigation, the British formed the InterService Ionosphere Bureau (ISIB) in 1941. The United States followed suit in 1942 by establishing the IRPL at the National Bureau of Standards (NBS) in Washington, D.C., with military funding. Its purpose was to collect and disseminate the ionospheric data taken by scientists and military units stationed around the globe. The organization began to issue daily condition reports that predicted radio transmission quality for the next seven days. These radio quality conditions were rated 1 to 9 (1 meaning useless reception, 9 meaning excellent reception) for each of three regions: polar-auroral, midlatitude, and equatorial. Because of airplane navigation problems, the IRPL issued another simple daily message for pilots: Warning (W) or No Warning (N). The W and N messages were sent worldwide.

The commercial sector aided the war effort by contributing to these simple forecasts and also benefited from them. RCA and AT&T agreed to alert staff at the IRPL when they had unusual recordings on their ground-current devices, or when transatlantic messages dipped below commercial quality. Immediate alerts from these two companies, day or night, allowed for quick dissemination of high-quality information. Timely warnings of disturbances proved invaluable to radio operators, who often wrongly assumed that disruption in a signal was due to equipment failure. Knowing that a disturbance was caused by space weather saved them time looking for a malfunction in transmitters and receivers.

These first attempts at systematic forecasts of radio disturbances proved remarkably accurate. Like terrestrial weather forecasts, space weather forecasts were based on two initial principles (Fig. 2.3). One was persistence: the concept that the weather will continue to be the same. Solar storms, once detected, seemed to have a persistence that would last for at least several days because the active regions on the surface of the Sun lingered, changing in shape and character. From 1942 to 1944, the persistence forecasts often lasted for about five days. The second principle was recurrence: the concept that an event will happen again at a predictable time. This principle was based on the twenty-seven-day rotation period of the Sun. If, just before rotating out of view, the Sun's west limb (the right side as you look up at the Sun)

Figure 2.3: *These diagrams illustrate the principles of recurrence and persistence as they relate to the Sun. The orange circles represent the Sun as it rotates; the dark spots represent active regions. (NOAA)*

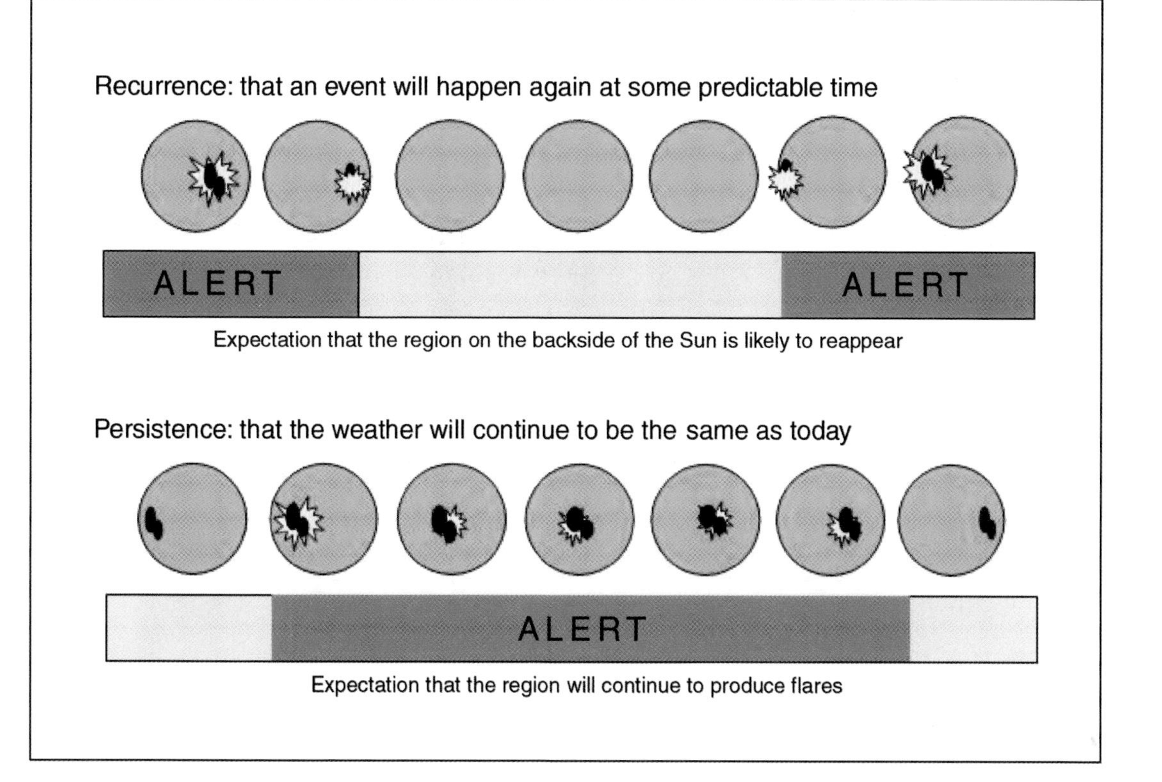

showed a very bright area, it might well reappear on the east limb (left side) in about fourteen days when that area rotated back into view. These bright areas typically indicated solar activity, likely to create a geomagnetic storm that would affect Earth. An active region on the west limb would likely still be active when it reached the east limb, and observers braced for a geomagnetic storm several days after an active region appeared. During this phase of Solar Cycle 17, in the early 1940s, these two principles led to remarkably accurate forecasts of radio disturbances.

These techniques were not based upon a complete understanding of the Sun's activity. In fact, forecasters in the 1940s got lucky. During times of high solar activity (Solar Maximum), as in the early 1940s, the principles of persistence and recurrence are more reliable. The situation in terrestrial weather is similar. For example, during the hurricane season, meteorologists can safely guess that if it rains today, it will rain tomorrow (persistence), and if there is a hurricane in Jamaica today, there will be stormy weather in Florida in a few days (recurrence). Although observers could see sunspots and solar flares (two types of visible activity that can lead to space weather storms), people interested in forecasting radio disturbances did not necessarily know why their predictions were accurate. As observers would later discover, visible solar activity is highly correlated to space weather storms, but it is not necessarily the cause. During times of low solar activity, the Earth can still be affected by the Sun.

By February 1944, the lucky streak of radio disturbance forecasting began to run out. When the Sun's seasons began to change, the principle of recurrence for forecasting became unreliable, and coronal observations were not a good diagnostic method for forecasting. Scientists from the United Kingdom and United States tried to understand the change based on the little they knew about recurring events. Although they could no longer make long-range predictions about radio quality, they could send out HF condition alerts a few hours in advance of disturbances. These alerts merely reported current conditions "in real time," that is, as soon as conditions could be measured. The recording magnetometer located at the NBS field station at Sterling, Virginia, proved vital in providing these data.

With the acknowledged importance of terrestrial weather forecasting in war, the U.S. military was interested in finding out whether solar disturbances, especially the auroral ones, had an impact on terrestrial weather. Birkeland too had grappled with this idea in his time. Weathermen during the war recognized that if the aurora brought on, or otherwise influenced, terrestrial weather, then weather forecasting would be improved with observations of the aurora. In the early 1940s,

British, American, Russian, and German weathermen independently set up weather stations across Greenland, Canada, and Scandinavia. For the most accurate weather forecasts, these stations were established close to the theater of war, and incidentally high enough in latitude to permit excellent aurora viewing. In 1946 a large number of solar storms excited tremendous auroras, leading to what the scientists thought were "accompanying" weather storms.

The military was hardly the only group interested in terrestrial and space weather correlations. As the war ended, a group of midwestern farmers expressed their need for better weather forecasts for more efficient and lucrative farming. Away from the cities, they would occasionally see auroras over their dark skies, and hoped that the aurora would be a good weather forecasting tool. These farmers convinced their congressional representatives to allocate money to the Department of Agriculture to investigate a possible relationship between solar activity and Earth weather. The department turned to Harvard University for help. Harlow Shapley, the longtime head of the Harvard College Observatory, believed from his observations that the solar corona was affecting Earth's weather. In 1940 he provided funding for the young veteran turned graduate student Walt Roberts to install a solar coronagraph station at an astronomical observatory in the high, clear skies of Climax, Colorado, under the direction of Don Menzel, his thesis adviser at Harvard.

Menzel was a devotee of the French scientist Bernard Lyot, who had invented the coronagraph. This instrument allowed the observer to see the outermost layer of the Sun (the corona) by creating an artificial eclipse—an occulting disk was used to cover the Sun. The resulting image showed the Sun's atmosphere (the coronal and solar prominences). Menzel found coronagraphs intriguing and wanted to set one up in his boyhood home in the Rockies of Colorado. He managed to persuade the Climax Molybdenum Company to allow Harvard to build an observatory .4 km (.25 mile) from the mine, directly on top of the pollution-free Continental Divide. Roberts agreed to set up the observatory and live there for a year (which turned into seven), beginning in 1940.

Before departing for Climax, Roberts and his bride, Janet, married in New Jersey. The local Lakewood newspaper, in an advance announcement of the wedding, reported that the young couple would be studying the Sun and that "it is expected that the findings will have an immediate practical importance for the forecasting of such electrical disturbances as that which crippled the world's communication services on Easter Sunday [1940]." The lengthy article went on to discuss other possible space weather events that Roberts's studies might elucidate, such as "why [Boston newspapers'] wire services had relayed jumbled messages and why the sky was brightened by a peculiar display of what was thought to have been the aurora borealis." Whether or not the event correlations were accurate, the article contained an impressive amount of scientific information for a wedding announcement.

The newlyweds moved into a small house attached to the observatory that was perched up on the Divide, 11,500 feet above sea level (Fig. 2.4). As Roberts set up the coronagraph, the good-natured couple struggled through seven feet of snow, high-altitude baking disasters, and isolation. Walt joked that Janet, ever environmentally conscientious, would grow concerned when the news reported drought around the country: if there was a drought in the East, she would dump her washing water on the east side of the mountain. She was equally considerate of the people in the West.

During the war, Roberts observed the Sun and eventually gained an array of assistants. On June 4, 1946, one assistant urgently called Roberts to come see an enormous prominence. To date, this prominence remains the largest ever photographed. The experience and knowledge Roberts gained at Climax would launch him on a lifelong career in astrophysics. After leaving Climax for Boulder, Colorado, he became interested in the connection between weather and the aurora.

Figure 2.4: *The Climax Observatory was home to the coronagraph established and run by Dr. Walter Orr Roberts in 1940–1947. The house in which he and his family lived was behind the observatory and sat exactly on the Continental Divide, 11,500 feet above sea level, at Fremont Pass, Colorado. (NCAR)*

Walt Roberts joined forces with the scientist Roger Olson in the mid-1960s to try to identify a link between solar events and Earth weather. Olson had served as a weatherman for the Air Weather Service in Goose Bay, Labrador, during the war. Roberts and Olson attempted to convince the space weather community of a relationship between sunspots or solar flares and terrestrial weather. Ultimately, the definitive results that they worked for eluded them. To date, a cause-and-effect connection between space weather and terrestrial weather is still in question. Roberts later admitted that much of the research done in the field, including his, was "poorly done," "with sloppy statistical methods," and the "results grossly over interpreted." These criticisms implied that he believed his theory still *could* be correctly proven. In a retrospective paper presented on the thirtieth anniversary of the Office of Navy Research in 1976, Roberts said he felt as if the answers were just around the corner but admitted that there was still widespread skepticism among many, if not most, of the leaders and pacesetters in meteorology.

The war had broadened the interest of the scientific community concerning the problems of space weather, and it also fostered an energetic movement of international groups that wanted to pursue scientific issues as they related to *all* countries. The need for collective data that could paint a holistic picture of solar effects around the world, regardless of inclement weather or time of day, was at the core of such international cooperation. The number of countries collaborating on space weather research has grown to an impressive level, and international organizations continue to influence political decisions and ease international tensions. Much of what is done with modern forecasting of space weather has been formed on a history of international cooperation that continues today.

3 International and National Cooperation

The massive highway system that stretches across the United States could never have been built by one person. The tremendous cost aside, one person could never bring about the kind of cooperation needed between the states to complete such a project. It would take a powerful organizing body—the federal government—to foster the required cooperation. Space weather research, like the interstate highway system, could never have been undertaken by only one organization. Solar activity affects too many human systems and too great a geographic area for one organization to handle. The second half of the twentieth century would see important partnerships develop among countries around the world, even ones divided over the war, and among agencies in the United States. These partnerships led to massive projects, undertakings too great to be handled by just one country or organization. Space weather services would have stayed in their infancy without cooperation.

It was obvious nearly from the start of space weather forecasting that problems with technology and lack of scientific knowledge reached far beyond national shores. After the war, a mini–global economy existed. Countries needed to communicate with each other, to safely control airplane travel, and to work out global issues involving trade and commerce—all necessities that have components of space weather hazards. In order to communicate between countries, there had to be standards, such as the assignment of radio wavelengths for specific users. During the age of observational satellites that constantly circle the globe, international cooperation proved necessary for tracking satellites and recording and sharing data.

Several international science and policy organizations were formed that would become the basis for exchange of data and products throughout the world. Each group also took on the responsibility for advising on policy relating to the global issue on which it specifically focused. For example, the Union Radio-Scientifique Internationale advises on matters dealing with international problems with radio. These groups proved vital because no one body could successfully address all of the many global issues. The history of these organizations, each referred to by an acronym, can be confusing. Some were formed by the merging of specifically focused organizations; others split off from parent organizations and turned into committees or "unions"; still others began on their own and later joined newly formed superorganizations. Together, these key organizations are responsible for the development of international cooperation.

International Agencies

International Council for Science

The International Council for Science (ICSU, originally called the International Council of Scientific Unions) was founded in 1931 to promote both international activity in the different branches of science and the application of science for the benefit of humanity. It is a nongovernmental organization whose roots reach back to 1899. ICSU's strength and uniqueness lie in its global membership, which includes both national scientific bodies (103) and international scientific unions (27 that meet in international groups). This wide spectrum of scientific expertise allows ICSU to have a weighty standing when addressing major international interdisciplinary issues. ICSU has participated with many countries and the United Nations in addressing such global issues as the environment, sustainable development, and information technology. Most recently, ICSU has focused scientific and technological intervention on realizing sustainable development projects in Africa. ICSU also has concentrated on disaster reduction and coordinated research on the Earth's systems, mapping hazard areas, models of disaster scenarios, warning systems, and quick dissemination of information.

Space weather research is one part of the vital service ICSU contributes to international science. One of ICSU's bodies, the Committee on Space Research (COSPAR), was established in 1958 as an interdisciplinary scientific body concerned with the progress on an international scale of all kinds of scientific investigations that could exploit the vast possibilities of space vehicles, rockets, and balloons. The Scientific Committee on Solar-Terrestrial Physics (SCOSTEP) also keeps an eye on space weather. Principally it promotes, organizes, and coordinates international interdisciplinary programs in solar-terrestrial physics. Other members of ICSU relevant to space weather and services are included in the impressive roster of prestigious international groups that coordinate through the superorganization of ICSU (Fig. 3.1).

International Union of Geodesy and Geophysics

Established in 1919, the International Union of Geodesy and Geophysics (IUGG) became an original member of ICSU in 1931. Its

International Council for Science (ICSU)

76 National Scientific Members
29 International Scientific Unions, including:
Union Radio–Scientifique Internationale (**URSI**)
International Union of Geodesy and Geophysics (**IUGG**)
20 Interdisciplinary Bodies, including:
Committee on Space Research (COSPAR)
Scientific Committee on Solar-Terrestrial Physics (SCOSTEP)
Federation of Astronomical and Geophysical Data Analysis Service (**FAGS**)
and under FAGS, International Space Environment Services (**ISES**)
Panel on World Data Centres (Geophysical, Solar and Environmental) (**WDC**)

International Union of Geodesy and Geophysics (IUGG)

66 Adhering Bodies
7 Semiautonomous Associations, including:
International Association of Geomagnetism and Aeronomy (IAGA)

United Nations

United Nations Educational, Scientific and Cultural Organization (UNESCO)
World Meteorological Organization (**WMO**)

Figure 3.1: *These organizations (and many others) cooperate in each other's committees and programs. Although there appears to be a hierarchy, the structure does not imply order in terms of funding or precedence. The acronyms in bold are directly tied to SEC and are mentioned later. (NOAA)*

objectives are the promotion and coordination of physical, chemical, and mathematical studies of the Earth and its environment in space. The union is a federation of seven semiautonomous associations, each responsible for a specific topic or theme, including scientific studies of Earth's gravitational and magnetic fields, the ionosphere, the magnetosphere, and solar-terrestrial relations.

IUGG gives particular emphasis to the scientific needs of developing countries, and strives to improve the capacity of all nations of the world to observe and understand the natural physical processes that bear upon their safety and economies. JoAnn Joselyn, the current secretary-general, is a space physicist and a former space weather forecaster (Fig. 3.2). This background has helped her to lead the organization in its many-faceted interests.

Union Radio–Scientifique Internationale

Union Radio–Scientifique Internationale (URSI) first met in July 1922 in Brussels. At that time, only four nations participated in the union: Belgium, France, the United Kingdom, and the United States. But within the next year, Australia, Spain, Italy, Japan, and the Netherlands also formed committees as part of URSI. The group concerned itself with knowing about all aspects of radio science, a field witnessing fantastic growth at that time, and with extending the roles of radio

Figure 3.2: *Dr. JoAnn Joselyn, the secretary general of IUGG, fulfills her mission in supporting science collaboration around the world and focusing research on important world issues. Exemplary in many ways, she is a leader in her field and her community and has received many awards for her contributions. Joselyn worked at SEC for twenty years and continues to support space weather data exchange and research worldwide. (NOAA)*

and its electronic applications throughout Earth and the solar system. It was another ICSU charter member.

URSI members realized the importance of a detailed knowledge of the ionosphere for radio communications. In 1933 it observed radio disturbances from the Eiffel Tower and developed radio disturbance warnings (called URSIgrams), which it then transmitted from the tower. These warnings alerted radio listeners to expect interruptions of transmission and reception. Edward Appleton, a Nobel Prize winner recognized for his part in describing the role of the ionized atmosphere in transmitting radio waves, was president of URSI from 1934 to 1952. To this day URSI operates as a nonprofit, nongovernmental organization. It assumes responsibility for stimulating and coordinating, on an international basis, studies, research, applications, scientific exchange, and communication in the fields of radio science.

Federation of Astronomical and Geophysical Data Analysis Services, World Meteorological Organization, and World Data Centers

Whereas the above-mentioned groups focused on broad, general scientific research, several other international organizations focused more specifically on space weather data and alert exchanges. In 1956 ICSU formed the Federation of Astronomical and Geophysical Data Analysis Services, which was supported by IUGG. FAGS provided the scientific community with a long series of observational data, some going back to 1800. These past data came from the various contributing organizations. One of FAGS's current service organizations, the Sunspot Index Data Centre, was founded in 1981, but its database includes daily sunspot numbers since 1818 and yearly sunspot numbers since 1700. Another organization under FAGS is the International Space Environment Service (ISES, discussed in a later section).

The World Meteorological Organization (WMO), a United Nations body, likewise cooperated and consulted with ICSU, IUGG, and URSI and provided valuable weather data for these groups' studies of climate and weather. This weather organization has lately taken on the distribution of space weather products created by the Space Environment Center (SEC) as part of the distribution of WMO alerts, warnings, and watches worldwide.

The World Days Service and the World Data Centers (WDC) were initiated in 1959 as part of the International Geophysical Year (IGY, discussed in a later section). The World Days Service organized specific universal viewing days so that scientists all around the world could look at the Sun at the same time during a month and compare observations. The World Data Centers captured and archived those observations. The World Data Centers system now includes twelve countries that fund and maintain fifty-two centers on behalf of the international science community. Data collected at these centers encompass a wide range of subjects, including solar, geophysical, environmental, and human issues. These data cover timescales ranging from seconds (real-time data from satellites) to millennia (such as information taken from ice cores), and they provide a data library to aid research done by scientists working in the vast range of ICSU disciplines. All data held in WDCs are available for no more than the cost of copying and sending the requested information. The main U.S. World Data Center-A is housed at the National Oceanic and Atmospheric Administration (NOAA) National Geophysical Data Center in Boulder.

The International URSIgram and World Days Service (IUWDS) was formed in 1962 by a merging of the World Days Service and the former URSI Central Committee of URSIgrams (Fig. 3.3). IUWDS comprised a dozen Regional Warning Centers around the world that would exchange solar, geomagnetic, and ionospheric data each day. Each

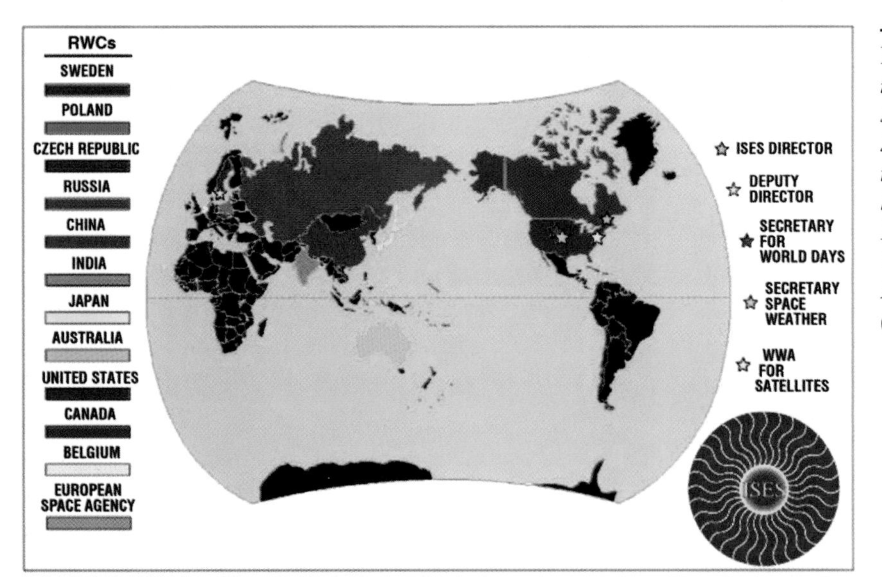

Figure 3.3: *Membership in the International Space Environment Service (ISES) continues to be strong since it began in 1958 as the International URSIgram and World Days Service (IUWDS). (ISES)*

regional center contributed its advice to the World Warning Agency, which would issue a consensus forecast of solar-geophysical activity for each day. Currently the organization has warning centers in Beijing, Boulder, Moscow, Paris, New Delhi, Ottawa, Prague, Tokyo, Sydney, and Warsaw. The Boulder Regional Center, at SEC, is the one World Warning Agency for the organization, as it has been since 1965.

Until IUWDS was formed, terse warnings and alerts of radio disturbances were sent once a day on weekdays. Dissemination of the alerts was slow and so not sufficient for real-time use (Fig. 3.4). Affected parties really needed a brief, standardized message that could be decoded and understood in any language. The Central Radio Propagation Laboratory (formerly Interservice Radio Propagation Laboratory [IRPL]) took on the responsibility of developing and maintaining the standard messages for space weather. In 1995 the name IUWDS was changed to the International Space Environment Service (ISES). Although "IUWDS" rolled off the tongue of those who used it every day, many people who used the service bashfully did not know what IUWDS meant.

International Research and Joint Projects

It is fair to say that the scientific community was well on its way to forming international cooperation even before the war, but World War II created a powerful incentive for international political partnership. The already strong cohesion among the Allies grew in light of the threat of Soviet expansionism. In 1948, France, Britain, the Netherlands, Luxembourg, and Belgium formed the Western Union to strengthen European defense. After another year of intense diplomatic consultations, the North Atlantic Treaty Organization (NATO) was formed in April 1949, comprising the Western Union, the United States, Canada, Denmark, Iceland, Italy, Norway, and Portugal. This international partnership initially centered on Western allies but grew into a global partnership that recognized the need for collaborative scientific research and information exchange.

In the climate of rebuilding and expansion after the war, international cooperation would help bring about one of the most important international research projects to date—the International Geophysical Year (IGY), which would focus a coordinated research effort on many physical aspects of Earth. A precedent had already been set for this type of project. The First International Polar Year (IPY), 1882–1883, had been inspired by a German arctic and antarctic explorer, Karl Weyprecht, who advocated internationally coordinated exploration of the polar regions. He thought scientific research should be the goal of polar expeditions rather than exploration for individual or national

Satellite Data Codes

UTELC

Content:

Total electron content.

Example:

UTELC	20401	80712	01525	51203	61304

Definition of symbols:

UTELC	IIIII	YMMDD	OHHmm	Habpp	[Habpp ...]

UTELC comes from Total ELectron Content

UTELC	**IIIII**	YMMDD	OHHmm	Habpp	[Habpp ...]

IIIII = station indicator (see lists in Appendix C)

UTELC	IIIII	**YMMDD**	OHHmm	Habpp	[Habpp ...]

Y = last digit of year
MM = month of year, 01 = January, 02 = February, *etc.*
DD = UT day of month

UTELC	IIIII	YMMDD	**OHHmm**	Habpp	[Habpp ...]

O = indicates that the begin time of the observation follows
HHmm = UT hour and minute of beginning of observation

UTELC	IIIII	YMMDD	OHHmm	**Habpp**	[Habpp ...]

H = last digit of UT hour
ab = total electron content in units of electrons/m^2 where **ab** = **a.b**
pp = power of ten to apply to **a.b**

Note: **abpp** *reported as* **1202** *would equal* 1.2 x 10^2 Wm^{-2}

Note: */ is to be used for data not available.*

Figure 3.4: *A page from the IUWDS code book explains how to decode satellite data messages. At the time, slow and limited technology meant that cursory messages were the only way to send data quickly and with regularity. (ISES)*

glory. Most early exploration brought fame and fortune or defeat and death, and scientific gains were not the first consideration—claiming the prize was. The fundamental concept behind the First IPY was that geophysical phenomena could not be surveyed by one nation alone; rather, an undertaking of this magnitude required a global effort. Coordinated and directed data had to be collected and shared among the countries of the world. Eleven countries participated in fifteen polar expeditions, and the IPY heralded a new age of scientific discovery rather than exploratory glory.

A Second International Polar Year was planned for the twenty-fifth anniversary of the First IPY at an international conference of meteorological service directors. This time forty nations participated in arctic research from 1932 to 1933. The focuses were the global implications of meteorology, magnetism, atmospheric science, and the "mapping" of ionospheric phenomena that would advance radio science and technology. Research had already shown that extreme ultraviolet light emitted from the Sun produces the Earth's ionospheric layers, and that these layers could be disrupted during magnetic storms. Ionospheric disturbances were much more pronounced in the polar regions, so scientists took equipment to Tromsö, Norway, to investigate how energetic charged particles cascade into the polar atmosphere after being guided there by the Earth's magnetic field. The Second IPY studies revealed an ionosphere that was more complex than had been previously known, one made up of various layers that each behaved differently. For example, in middle latitudes, it was shown that the maximum F2 critical frequency was affected by the equinoxes, spring and fall. On the other hand, the E-layer critical frequency changed with the position of the Sun. The D-region exhibited a seasonal anomaly in the high end of the middle latitudes. Sadly, the worldwide economic depression limited the Second IPY research studies. However, as a follow-on to the scientific studies from the Second IPY, a major loss of radio communications that occurred on May 12, 1935, was correlated with a solar flare. This well-documented event first established a link between a particular type of solar activity and a shortwave fade.

The Third IPY (1957–1958), later renamed the International Geophysical Year, was proposed in 1952 by ICSU, following a suggestion by Lloyd Berkner, a National Academy of Science member (Fig. 3.5). During that time, scientists knew that the solar cycle would be at its maximum, allowing them to observe the highly active Sun and conduct an extensive program of ionospheric research. Having scientists around the world observe the same activity at the same time, day and night, exponentially expanded the value of the observations. Simultaneous observations of the aurora at the north and south poles immediately verified the symmetrical nature of the phenomenon. Tracking ionospheric disturbances from several locations resulted in a three-dimensional picture of the disturbance.

In October 1954, ICSU, the organizing council for the IGY, adopted a resolution calling for "artificial satellites"—at the time, the word *satellite* was used for a celestial body orbiting another, such as the Moon

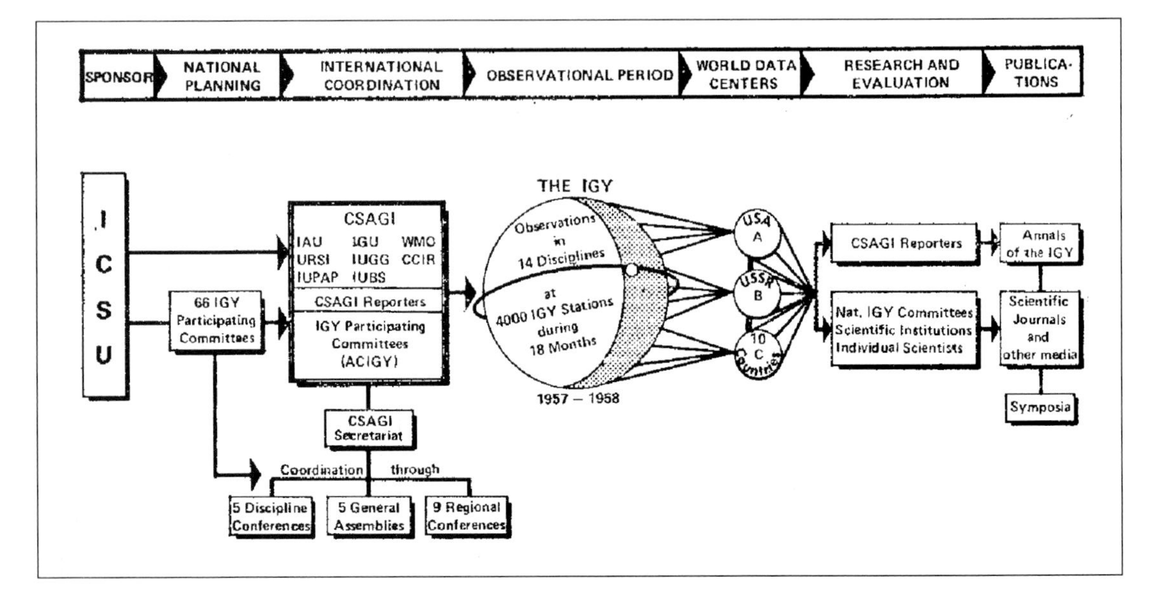

Figure 3.5: *How IGY explained its organization in* Department of State Bulletin 40, *May 11, 1959, p. 687.*

around the Earth—to be launched during the upcoming IGY to map the Earth's surface. Previous attempts had been made to look at the Earth-Sun connection with high-altitude rockets. These rockets, armed with data-collecting instruments, would soar into the upper atmosphere, collect a raft of data, than crash to Earth again. Satellites that could go into space and remain there were the next logical, desirable step.

The International Geophysical Year was "the largest, most complex, and most comprehensive international scientific undertaking thus far conceived and successfully carried out by scientists," reported the Subcommittee on National Security Policy and Scientific Developments of the Committee on Foreign Affairs, U.S. House of Representatives, November 1973. "Sixty-seven nationals participated, represented by 20,000 to 40,000 scientists and as many volunteer observers, [who] manned about 4,000 principal stations and an equal number of secondary stations and sites scattered throughout the world from pole to pole." Among the surprising scientific discoveries made during the IGY was the highly energetic radiation Van Allen Belts around the Earth, discovered by the U.S. satellite Explorer I.

In the 1950s politics and science became intertwined, combining the need for political and scientific cooperation. The Cold War threat of a nuclear war with the Soviet Union inspired efforts to monitor space weather events that might cause the United States to miss the launch of an attack. In response to the call for satellites for the IGY, President Dwight Eisenhower announced plans in July 1955 to launch an Earth-orbiting satellite for use during the IGY (mid-1957 and 1958) and solicited proposals from various U.S. government research agencies to undertake creating the satellite. In September 1955, the Naval Research

Laboratory's proposal for its Vanguard satellite was accepted. The proposal represented a contribution of major significance to the IGY by the United States. Vanguard was a 3.5-pound suite of science instruments (called a payload) strapped to a rocket that would propel the satellite into space. Vanguard was expected to be the first satellite, but the United States would not be the only country busy building a satellite.

As a follow-on to the IGY, scientists became interested in satellite-borne radar that could explore the ionosphere with topside sounding, which would be essential for communicating with satellites in space. Communication from the ground-to-ground stations had been well researched and the underbelly of the ionosphere "mapped." A topside sounder would allow for the study of transmissions from space-to-ground stations and the mapping of the top of the ionosphere. With the growing interest in satellites, this information was vital. Early in 1958 the Naval Research Laboratory invited Canada to work on a proposal to build a satellite that would carry a topside sounder. The Defence Research Telecommunications Establishment (DRTE) answered the call. DRTE, located in Ottawa, had its origins in ionospheric-sounding activities carried out by the Canadian military during World War II. By the late 1950s the DRTE primarily researched radio communications with its radiophysics and electronics laboratories.

DRTE wrote a proposal to build a topside-sounder satellite using a sweep frequency similar to the ionospheric sounders then in use in ground-based instruments. The sweep-frequency instruments would transmit and receive all of the possible frequencies to gather data, as opposed to fixed-frequency instruments that would send out only one frequency. The National Aeronautics and Space Administration (NASA) and the Central Radio Propagation Laboratory (CRPL) worked in close consultation with DRTE's talented team. This project proved to be large and ambitious; many Americans and even some Canadians thought it couldn't be done, partly because the Canadians had less satellite-building experience than the Americans. The American agencies were afraid that the plan was too ambitious and urged a reduction in its scope. Meanwhile, in order to hedge their bets and ensure the success of the topside-sounding project, the Americans were also building a fixed-frequency topside sounder.

Development of NASA's satellite suffered technical delays, but the Canadian satellite remained more or less on schedule. Contrary to expectation, the Canadians launched their first satellite on September 29, 1962. Alouette I was the first satellite to be designed and built by a nation other than the United States or the Soviet Union. The NASA satellite was eventually launched in August 1964. Colin Franklin, one of the original members of the Alouette program, noted, "NASA later admitted publicly that they and the CRPL ... were so convinced that Alouette could not possibly function for more than an hour or two, if at all, that they had made no plans to use data from it." A key to the successful Canadian design was a powerful transmitter that produced a signal strength ten times that of the calculated minimum needed and resulted in mass production of high-quality ionograms. Alouette I, constructed at a time when most satellites had a useful life span of a few months, continued to function and provide a wealth of data for ten years before being turned off from the ground. It was one of the most complex satellites of its day, and it set several precedents for technological discoveries. Canadian scientists added to their expertise and gained prominence as world experts on the upper atmosphere.

Despite vicious Cold War tension (discussed further in Chapter 4), the United States and the Soviets came together in the name of science. By the 1970s the Soviet space industry had taken a new direction in the manned space program—prolonged missions for men in space. In 1971 the Russian Soyuz shuttle delivered three cosmonauts to the first Soviet Earth-orbiting space laboratory, called Salyut 1. Tragically, the crew of Salyut 1 died while reentering the Earth's atmosphere. Addressing the dangers and difficulties of working in space superseded any national hostilities, and the two superpowers came together in 1975 to create the Apollo-Soyuz mission, using the Soviet Soyuz and the U.S. Apollo spacecrafts to dock with each other while in orbit (Fig. 3.6).

The Soyuz spacecraft, modified over the years, became the major transport to Salyut's successor, the Mir space station, during its construction between 1986 and 1996. In 1995 NASA formed the Shuttle-Mir Program with the Russian space agency, and the U.S. space shuttle docked at Mir for the first time. Scientists in both countries learned a great deal from the space station missions, especially in the area of long-duration space living: the Russian medical specialist Valery Polyakov spent a record 438 days in space. Some experiments outside the space station permitted study of the space environment, specifically how long-term exposure to space, with its vacuum and debris, affects spacecraft. The Russians guided Mir, a venerable and crippled craft by that time, down to a watery grave in March 2001. The unprecedented cooperation and trust between the U.S. and Russian space programs led to the development of the International Space Station in 1998. Although its future is in question, the International Space Station showed the spirit of continuing international cooperation in space research.

The European Space Agency (ESA, the European equivalent to NASA) was formed in 1975 and quickly became interested in sending a satellite to intercept Halley's Comet. Its Giotto satellite flew by the comet in 1986 and went on to complete other missions that explored

Figure 3.6: *This NASA commemorative design of the Apollo-Soyuz mission shows crew members from the United States and the Soviet Union as well as the two spacecraft. The mission signified a warming in the Cold War. Americans and Soviets used their own existing spaceships and engineered a compatible docking system. (NASA)*

Mars, the Moon, and Jupiter and its moon Titan. ESA and NASA joined together in 1995 to build and launch SOHO (described in Chapter 4), one of the most valuable satellites in operation today. ESA, along with Russia and Japan, also joined the United States in building the International Space Station. The coalition tested parts, contributed astronauts to mission crews, and assembled parts of the station. Like the SEC in the United States, ESA also provides space weather support to commercial groups in the European Union. ISES and the SEC offer support to ESA by sharing data and their experience of providing services.

In Japan, the National Space Development Agency (NASDA, now the Japan Aerospace Exploration Agency) has supported a vigorous space program since 1969. The rigorous program of missions includes Earth observation; communication, broadcast, positioning, and engineering experiments; upper-atmosphere/plasma observation; and lunar and planetary exploration. NASDA supported the building and launching of the space shuttle and the International Space Station, and Japanese astronauts have often served on the crews. Not only do the Japanese carry out superb research, but, as NASDA's website boasts, "the Tanegashima Space Center has a reputation as the most beautiful launch site in the world, and has been proud of this. NASDA felt that the reputation of its beauty reflected the feeling that 'Japan can do more than we thought'" (Fig. 3.7).

Figure 3.7: *From the Tanegashima Space Center, Japan launched the Yohkoh satellite. Japan's involvement in the space program has been vigorous and remains so to the present day. (Japanese Aerospace Exploration Agency)*

A Japanese partnership with NASA resulted in one of the pioneer satellites for space weather, which took its place as a Sun sentinel. The Japanese launched the Yohkoh satellite from the Tanegashima Space Center on August 31, 1991. The satellite, built in Japan, carried an observatory for studying x-rays and gamma rays from the Sun. The spacecraft was relatively novel because it orbited the Sun at the L1 point between the Earth and Sun, balanced by gravitational forces and orbital motion in such a way that it stays between the Earth and Sun at all times (as discussed in Chapter 4). This orbit is extremely valuable because the satellite is constantly looking at the Sun and is never in Earth's shadow; it is closer to the Sun than Earth by a million miles and reads the solar wind before it reaches Earth.

Yohkoh was tracked by Japan during that country's daylight hours and was left to collect and store data as Earth rotated around to the dark side. The x-ray imager carried by Yohkoh was built by Lockheed in California and had a tremendous vantage point from which to view the Sun. Although the images were extremely revealing and exciting, they were delayed in being relayed to the operational centers in real time because half the time, the data were held for up to twelve hours when the satellite was on the other side of the world. The information delivery was not quite timely enough for forecasting. Still, forecasters used the images whenever they could, and they learned much about the Sun. Yohkoh suffered its demise during a solar eclipse on December 14, 2001, when the satellite's direction finder lost its fix on the Sun. Without that orientation, Yohkoh could not aim the solar collectors at its power source and thus quickly ran down its solar power cells. Without power a satellite cannot send or receive signals, let alone carry out observations or run normal functions. Yohkoh had watched the Sun every day for ten years.

Although a TV broadcast aired around the world, many Americans failed to catch the remarkable launch of the first Chinese satellite. China's first manned spacecraft entered orbit on October 15, 2001, making it the third country to send a human into space. Chinese scientists, of course, eagerly checked the space weather forecasts at the SEC website in advance to ensure a safe and successful mission. The spacecraft, carrying a single astronaut, Yang Liwei, circled the planet fourteen times, returning to Earth after a flight of about twenty-two hours. Although Yang returned safely to Earth, a home in Sichuan Province was hit and damaged by the remnants of the craft. Although China reached this milestone without help from other countries, its interest in space shows that it is reaching out to the international community. Future partnerships between China and ESA seem likely.

International cooperation made possible the rapid development of satellites and space stations. These remarkable relationships continue to grow, as does the world's library of information about space and the Sun. When one considers that many of the partners in these efforts have, at one time or another, been powerful enemies, it seems remarkable that dedicated scientific cooperation has lifted them above political sparring.

Interagency Cooperation in the United States

In order to step up to such international commitments—and lead the way on many space and space weather fronts—cooperation among the agencies within the United States would be crucial after World War II. NASA, NOAA, and the U.S. Air Force (USAF) shared a mutual interest in forecasting space weather, and each organization could contribute a unique aspect to the research and monitoring. Because NASA needed to assess the solar and cosmic radiation risk to astronauts, especially when the astronauts ventured outside Earth's magnetospheric bubble, the agency planned to build and install ground-based telescopes. The Air Force was prepared to send up satellites in order to have excellent space weather monitoring for radio communications as well as for other technological military applications. NOAA contributed scientific expertise and services as well as trained personnel to operate the telescopes.

National Oceanic and Atmospheric Administration

Since its inception in 1903, the U.S. Department of Commerce (DoC) has been "home" to the government's space weather and radio propagation work. The DoC sits at the top of a hierarchy of organizations that encompasses space weather research. The National Bureau of Standards (NBS) was a major element in the DoC. IRPL was formed under the NBS and put in charge of monitoring ionospheric disturbances for radio communication. In 1945, IRPL's work was declassified and its name changed to the Central Radio Propagation Laboratory, but the work remained the same. In 1954 the Department of Commerce chose to move some of its agencies, including CRPL, outside the Washington, D.C., area. The government built a large facility for the laboratories in Boulder, Colorado, and it was dedicated by President Eisenhower in September 1954. He termed the work to be done there a "new type of frontier ... bringing promise of a better, richer life, the opportunity to grow intellectually and spiritually ... as a people more happy and prosperous."

The move to Boulder would bring together several principal characters in space weather research and would foster interagency cooperation. Alan H. Shapley (Harlow Shapley's son) and Virginia Lincoln (a physicist who until 1942 had taught "household equipment" courses at Iowa State University) would dedicate most of their professional lives to CRPL, beginning their careers with launching the first systematic issuance of reports on ionospheric conditions. Lincoln worked on atmospheric research and began preparing monthly ionospheric prediction contour maps (which showed the density of the ionosphere) for

Figure 3.8: *Virginia Lincoln (right) was considered one of the greats in the field of space weather, having been chief of the solar-terrestrial physics division and director of the World Data Center–A. Early in her career she was the first woman to head a section in the federal bureau and was the only woman in the official U.S. delegation to the IGY. In this mid-1970s photograph she stands over a technician scaling ionograms. (NOAA)*

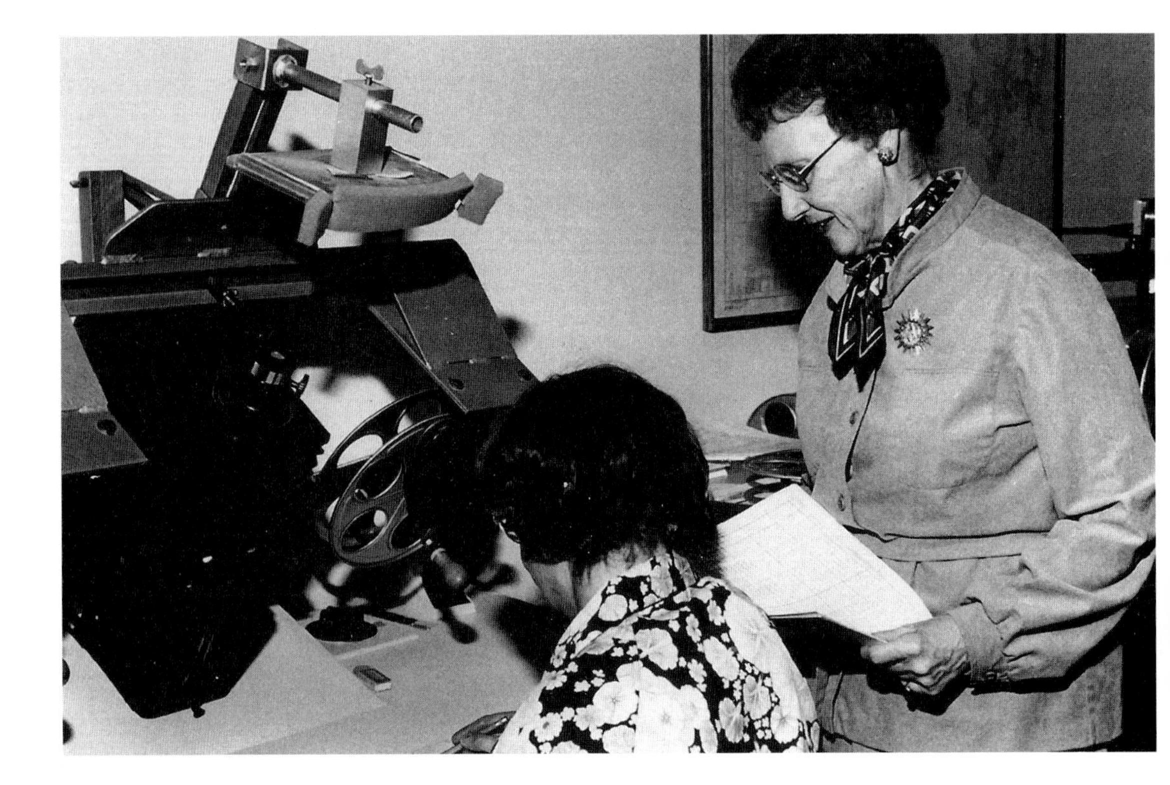

use in radio disturbance forecasting. In 1959 she was appointed chief of the Radio Warning Service and was the first woman to lead a section in the federal bureau (Fig. 3.8). Shapley, engaged in radio physics research, would come to work closely with Walt Roberts, who was now living in Boulder after his seven years high in the Colorado Rockies. Roberts had become the director of the High Altitude Observatory, at that time part of the University of Colorado. The prediction and alerting group under Virginia Lincoln and the strong research group in the Radio Physics Division, led by Alan Shapley, worked weekly with Walt Roberts to discuss solar-terrestrial relationships. They continued their work on radio propagation and rocket and missile tracking through the 1960s (discussed further in Chapter 4).

When people in the government explain the history of their organization, it can sound like a jumble of unknown languages. For example, the history of SEC sounds something like this: CRPL merged with the Weather Bureau and Coast and Geodetic Survey to form ESSA in 1963; as part of ESSA, the SDL and the forecasting center called SDFC held all responsibility for solar forecasting; ESSA split and part of it became NOAA in 1970; SDL became SEL, and SDFC became SESC, and the name change came with a slight change in job description; SEL became SEC, as it remains today (although SEC has moved into the NWS, still within NOAA). An attempt to translate this language is made in Chapter 8. Throughout all of the renaming and reorganization during the 1970s and 1980s, the space weather organization was led by two individuals who would push for different outcomes: one would promote exciting new science and one would bring about closer cooperation with another governmental agency.

These individuals, Don Williams and Bob Doeker, created a schism between research and forecasting in the space weather organization (Fig. 3.9). Williams, head of the Space Environment Laboratory (SEL), wanted to create the best magnetospheric research center (as opposed to one focused on the ionosphere) in the country. The forecasting arm of SEL, the Space Environment Services Center (SESC), took a different tack under Doeker's leadership. An ex–Air Force man, he knew that the Air Force needed mainly ionospheric products and services. He sought and won strong monetary support from the Air Force and began the very close relationship between the two agencies, formalized by an agreement between the Air Force and SEC. Since this relationship was established, there have always been a few Air Force personnel assigned to help staff the SESC. In this partnership the SESC provided the solar forecasts and information about the Sun to the Air Force, which in turn focused its research and forecasting on the ionosphere. Perhaps the best part of this arrangement was that the Air Force could not pass its data

Figure 3.9: *Bob Doeker (center left) and Don Williams (center) negotiated with the U.S. Air Force for a permanent cooperative agreement between the two groups. To this day the USAF and NOAA share the duties of monitoring and forecasting space weather. The results of the collaboration are distributed to civilian customers by NOAA and to the military by the USAF. (NOAA)*

directly to civilian users because of rules about classified information, but it could share data with SESC, which could pass it along to civilians through its products and services. The reverse is also true: the military could not receive critical space weather data from other countries such as the USSR, whereas SESC as a civilian agency could. Therefore, regardless of tensions between foreign governments and the United States, data sharing was never interrupted. In the end, the two directors pulled their groups apart, separating the researchers from the forecasters and setting up a "house divided" (further discussed in Chapter 8).

U.S. Air Force

The history of the Air Force is nearly as complicated as the history of SEC. The U.S. Army originally carried the military responsibility for monitoring weather. In 1947 the government created the U.S. Air Force out of a branch of the army, and responsibility for providing weather service to all the military agencies fell to the Air Force. Space weather monitoring and forecasting grew in the Air Force as it did in the civilian sector. The Air Force Weather Agency (AFWA) was formed on October 15, 1997, as part of a reengineering effort to streamline and improve the structure of the older Air Weather Service.

The Air Force discovered after World War II that the persistence model of forecasting radio disturbances no longer worked as it had in the early 1940s, owing to the changing solar cycle. It needed better ionospheric data and real-time forecasts. The Cold War increased the urgency of the need for accurate information about the ionosphere. Aware of the threat of nuclear warheads and intercontinental ballistic missiles, the Air Force joined with Canada to form the North American Aerospace Defense Command (NORAD). NORAD set up an incredible control and command center at Cheyenne Mountain, just southwest of Colorado Springs, Colorado. From this location, all very-high-altitude traffic could be monitored, including nuclear detonations, missile movement, and spy planes. The complex, built deep under the mountain, had every amenity to protect it from nuclear attack, including two 3.5-foot-thick solid steel doors weighing twenty-five tons each and armed with impressive hydraulic locks (Fig. 3.10). Work began at "the Mountain" as soon as construction was complete in 1966.

Along with NORAD, the complex housed the Space Defense Operations Center, which would rely on the support of the Space Disturbances Laboratory for its defense missions. As the DoC, not the Department of Defense, had been given the central federal responsibility for space weather forecasting, the Air Force would need the help of civilian government agencies. Air Force officials at Cheyenne Mountain called for meetings with CRPL in Boulder to gain knowledge about space weather disturbances. What, they wondered, would a space weather event look like on their systems? What were the best data to warn them of solar events? For CRPL staff, trips to the Mountain were understandably thrilling—the razor wire surrounding the massive dark tunnel entrance to the complex; the guards armed with loaded, fully automatic weapons; the secrecy surrounding pictures and documents; even the infrastructure of the Mountain, including the heightened security that could be backed up with "deadly force."

The Air Force Space Command (also part of the Space Defense Operations Center at Cheyenne Mountain) was created in 1982 to support satellite and space shuttle operations. In addition to keeping an eye on weather and communications satellites, a vital part of its mission is to track space debris. Since the launch of Sputnik (the first human-made object to enter space), the center has cataloged 26,000 objects that have gone into space. Most of these objects reentered the atmosphere and returned to Earth. About 8,500 remain in orbit, and these thousands of fragments of rockets, jettisoned trash, and broken satellite

Figure 3.10: *(Top) The entrance to the Cheyenne Mountain facility is small in the lower right of this photograph, but the facility inside the mountain is anything but small—or cozy! (Bottom) The blast-proof entrance doors leading to the heart of the facility are 3.5 feet thick and weigh 25 tons. (USAF)*

parts must be monitored (Fig. 3.11). Most are large enough to cause serious damage to functioning satellites. When a piece of junk is anticipated to float into the path of the Hubble Space Telescope at orbital speeds (perhaps 10 km per second, or six miles per second), Space Command advises NASA to move the orbiting telescope out of the way. As cumbersome as that may seem, a hit from a soda can could cause fatal damage to the Hubble. The Space Control Center also needs space weather information. When solar storms disturb the low-Earth orbits that Hubble and the space shuttle occupy, both the satellites and the space debris move differently. The storms heat the upper atmosphere, causing it to expand into the orbits of the objects. The objects then have to travel through the denser atmosphere, which slows their speed, potentially resulting in an object plummeting to Earth. Hence, Space Command must adjust its calculations for the orbits of the objects. Both NASA and NOAA rely on this timely information in order to protect their valuable investments orbiting in space.

Perhaps the most impressive cooperation between the agencies is the flow of talented people from one agency to another. Many of the people in the field of space weather worked for civilian government organizations, and many were retired Air Force men. In 1965 the USAF Air Weather Service began to send uniformed military staff to assist the Space Disturbances Forecast Center (SDFC) and to jointly provide the nation's space weather services. This cooperation increased until by 1972, some Air Force staff had been assigned to work long-term at SESC (previously SDFC), and the organization became officially jointly staffed. With the combined military-civilian knowledge of high-frequency communications and radar tracking of rocketry, satellites,

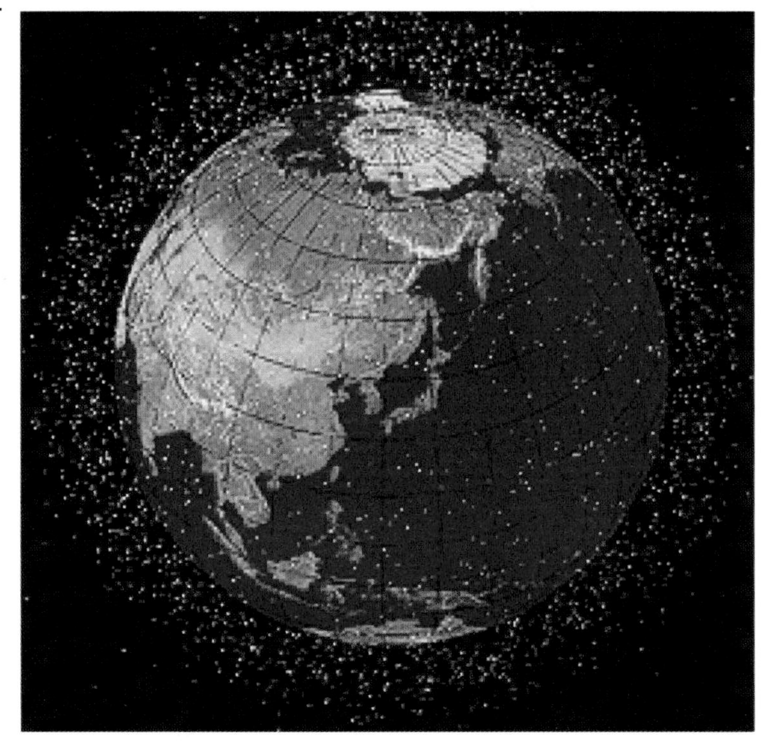

Figure 3.11: *In this plot of the debris orbiting the Earth, each dot represents a significant piece of "junk" big enough to cause substantial damage to a satellite. USAF Space Command tracks each of the thousands of pieces so that operators such as NASA can move valuable assets like the Hubble Space Observatory out of the way. Space weather can cause any of these debris objects to shift in their orbits; Space Command must recalculate their locations after every significant space weather storm. (NASA)*

and Global Positioning System (GPS), the SEC Forecast Center has been able to continually improve forecasting and serve an international community.

National Aeronautics and Space Administration

The National Aeronautics and Space Administration (NASA), established by Congress in 1958, inherited the earlier National Advisory Committee for Aeronautics and a few fragments of other government organizations. From the beginning, NASA began building on previous rocket and satellite technology to achieve the ultimate goal of human spaceflight and space exploration. As its scientists began deploying data-collecting satellites and running manned flights, they discovered that their missions were also being affected by space weather. Rocket and missile launches, satellites in orbit, astronauts with radiation-sensitive bodies, and radio communication could all be harmed by solar events. NOAA (and its predecessors) would interact with NASA groups on all of these problems and more.

In 1966 parts of the Environmental Science Service Agency (ESSA) stepped forward to provide support for NASA's manned Apollo Program. During the tense and challenging years of the mid-1960s, staff at SDFC worked shifts that required them to travel to Houston to interpret space weather data, oversee forecasting, and bring expert knowledge to support the missions. The staff also helped to run NASA's solar observatories around the world (discussed more in Chapter 4). Fortuitously, as officers from the Coast and Geodetic Survey (now the NOAA Corps) lost their jobs owing to funding problems in their own agency, they were assigned (some say "cast off") to SDFC, which needed the valuable services they could offer. These men and women also rotated through the Manned Spacecraft Center, named Johnson Space Center in 1973. The SDFC staff in Houston became known as "NOAA-Solar" to distinguish it from the NOAA National Weather Service also serving the Manned Spacecraft Center.

In the mid-1960s NASA funded the ground-based network of solar observatories collectively known as the Solar Particle Alert Network (described in Chapter 4). NOAA and the USAF staff operated these sites around the world and provided the data upon which SDFC based space weather alert predictions. Originally, NASA considered establishing its own in-house space weather operations center rather than asking NOAA or the USAF for help. In an attempt to be self-sufficient in terms of solar forecasting, NASA hired a contractor to develop a statistical model for predicting the size of solar storms. NASA then conducted a full manned-mission simulation using the model: everything running as it might during a real solar radiation storm, astronauts on the launchpad, computers running, countdown, and so forth. Within the framework of the simulation, the model predicted that such a (simulated) storm would last a year! Although it was a ludicrous prediction, NASA flight rule required a complete shutdown of the simulation and the launch in order to evaluate the results. This persuaded NASA managers that they did not want to take on the entire responsibility of predicting space weather. SDFC (at this point SESC) became the principal adviser to NASA on space weather, and the two organizations signed a formal support agreement in 1971. A simulation model developed by SESC made sure that radiation storm forecasts stayed within reasonable limits (in this case hours, not days).

All three agencies developed their own satellites, NASA and the USAF first, then NOAA. All three assisted in running the ground-based solar observatories. All three exchanged information and data and relied on the others for support during various projects. Today NOAA and the USAF run operational centers that issue forecasts, alerts, and warnings; NASA and the USAF launch most of the satellites; and NOAA leads the development of new space environment instruments that are built by NASA. Perhaps best of all, all three agencies provide services and support to other groups in the public and commercial sectors. For example, NASA has helped bring about a new generation of commercial communications satellites such as the Echo, Telstar, and Syncom satellites. USAF uses its data to support many "users" in all branches of the military and has owned and run the GPS satellites heavily used by the private sector. Finally, NOAA provides space weather services to support a huge variety of groups, from power companies to pigeon racers (discussed more in Chapter 7).

Cooperation among countries and between U.S. agencies spurred technological developments, research breakthroughs, and interest in space weather. Without these organizations, space weather research and forecasting would be limited at best. But as cooperation fostered scientific development, its antithesis—competition—also brought about impressive advances in the science of space weather. The Cold War, a political and emotional struggle between the Soviet Union and the United States, perhaps pushed scientific research further than cooperation could have.

The Space Race

The Cold War and the Space Race

For Americans, the relief of having won the peace in 1945 slowly turned to fear as tension between the United States and the Soviet Union grew into the Cold War. The frightened people of the United States built fallout shelters and amply stocked them with jars of food; they practiced "duck and cover" in schools to protect their heads from a nuclear blast (as if hiding under a school desk would help); they mistrusted everything "Red"; and they strove to do everything the Soviets could, only better. Americans developed entire organizations and programs to monitor nuclear testing by the Soviets. Both Soviets and Americans wanted to prove that they were the supreme technological power, and what better way to show supremacy than to conquer space? But as the Cold War grew, the race for space began to change almost imperceptibly. Fear turned to fascination, competition turned to cooperation, and racing turned to researching. Interest in exploring space and "conquering it" grew, and the technology went from satellites to manned flights in space to space-based observatories and orbiting stations. The Space Race introduced U.S. scientists to space and gave them invaluable knowledge about how to maneuver in space and, ultimately, how to use it.

The technology that would allow humankind to capitalize on space, and perhaps the most significant result of the Space Race, was the data-collecting satellite that could communicate with Earth. Americans learned that space-based technology could watch the clouds move over the planet, making possible the weather forecasting that we see on the evening news. Satellites located in a geosynchronous orbit made national TV broadcasts simultaneous and in real time, not to mention making worldwide telephone calling easy. The Global Positioning System (GPS) technology so widely used today allows for the use of extremely accurate location finding. These are all aspects of our lives today that we can't imagine living without.

The tensions of the Cold War centered largely on missiles and nuclear weapons. The Soviets had started aboveground testing of nuclear weapons, the second nation in the world to do so, on August

29, 1949. They detonated their nuclear devices mostly on Russian soil and in the Arctic Ocean. Annihilation of the United States seemed frighteningly possible. Therefore, detection of nuclear explosions became a matter of national security.

Scientists already knew what a nuclear explosion "looked like," thanks to the early tests carried out by the United States. They had collected a large number of patterns of data, or "signatures," characteristic of nuclear explosions. The Advanced Research Projects Agency (ARPA, later renamed D[efense]ARPA) started Project Vela, involving several techniques for detecting atmospheric nuclear explosions. The hope was that the presence of certain unusual elements or the number of nuclear-characteristic particles at high altitudes (at least 20 km [12.4 miles] above Earth) could indicate nuclear tests, thereby making enforcement of a test-ban treaty possible. The residue produced by nuclear explosions affected the Earth in much the same way that space weather does. Scientists could measure the atomic signatures by measuring radio propagation disturbances, alterations in the geomagnetic field, and electromagnetic pulses from the explosion in radio frequencies. Understandably, scientists and politicians needed to know if the atmospheric signatures were caused by a nuclear blast or by the Sun.

In 1962 ARPA gave its High Altitude Nuclear Detection Studies (HANDS) project to the Central Radio Propagation Laboratory (CRPL) because of CRPL's well-known experience with measuring the signatures of space weather—especially long-distance radio propagation. Scientists at CRPL set out to find the normal background of the atmosphere so that they could distinguish a nuclear event from the background. Of course, the atmosphere is far from consistently normal, so CRPL also tried to establish the "normal" background during various solar events. The HANDS project leader was Glenn Jean, who would later assist in the administration of the Space Environment Laboratory (SEL). He brought together a team of scientists and engineers to work on the project and, in particular, to take measurements, record data, and process the data with computers. The computers could collect data quickly (at rates of a tenth of a second), distinguish a natural event from a nuclear one, run twenty-four hours a day, and process long lists of data into plots that could be easily analyzed by the staff. In 1963 the computer used was built by Signal Data Systems (SDS, later bought by Xerox Corp.) and named XDS 930 (Fig. 4.1). The computer was massive and was generally run only when operators were present. The CRPL engineers were so bold as to run it over the weekend without a babysitter—perhaps the first ones to do so with this infant technology.

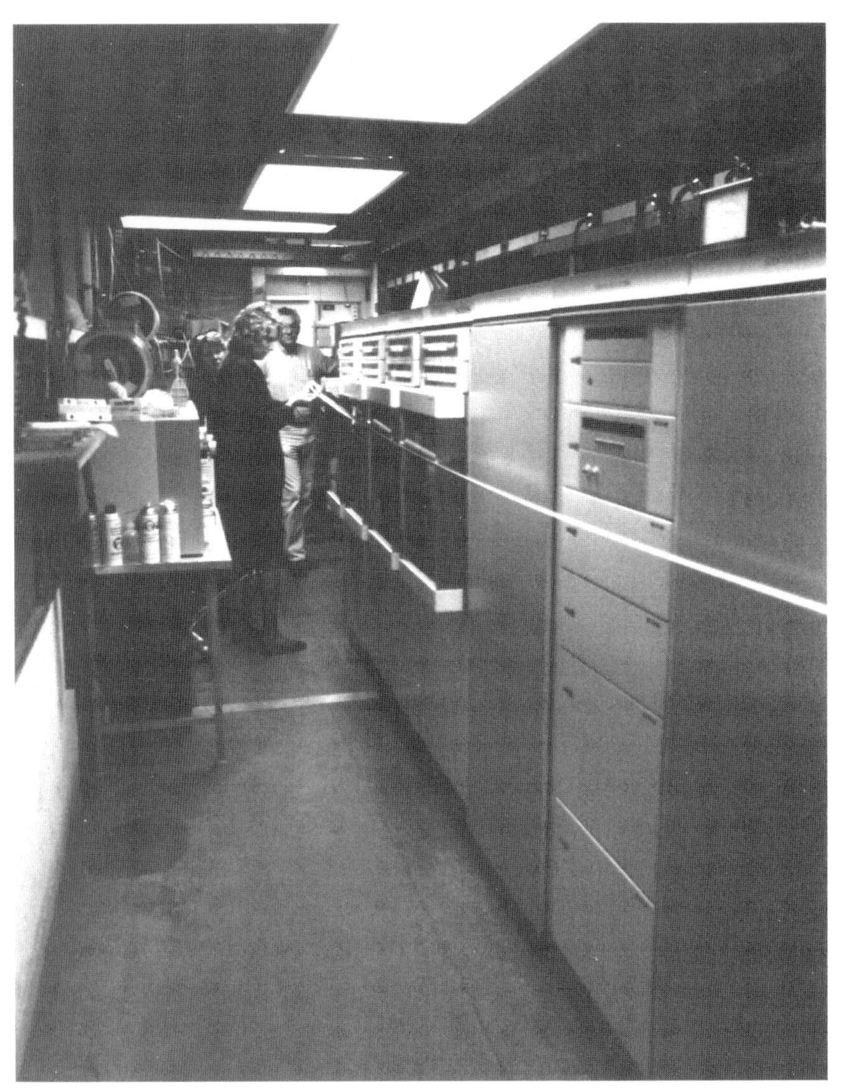

Figure 4.1: *The XDS 930 HANDS computer, being measured here in 1995 for shipment to a museum in California, contributed to space weather analysis between 1964 and 1970. (NOAA)*

Though looking for atmospheric data in the normal background range, the HANDS scientists incidentally observed numerous tests in the South Pacific. The U.S. high-altitude series of atmospheric nuclear tests from Johnson Island occurred in 1962. During that year, the United States dropped from planes or launched on rockets more than thirty-five bombs. This thermonuclear frenzy of superpower saber-rattling was the end of U.S. and Soviet atmospheric detonations. From December 26, 1962, to March 14, 1964, the USSR held to a moratorium of testing while the Atmospheric Test Ban Treaty was being negotiated; the treaty was signed on August 5, 1963. All further testing by the USSR and the United States was done underground. The HANDS team continued to monitor the atmosphere as other countries carried out atmospheric nuclear tests. The French began trying out their new atomic test site at the Moruroa atoll on July 2, 1966. Between 1966 and

1974, HANDS recorded forty-four French bomb detonations in the South Pacific.

The origins of the Space Race go back to World War II. During the war Germany was intensely interested in rocketry as a way to deliver bombs from a great distance. Not surprisingly, U.S. officials also took an interest in rocketry, partly owing to the German scientists who fled to America in the early days of Hitler's rise to power. Those German scientists helped the United States to explore its other main interest—the possibility of building an atomic bomb. The United States beat the rest of the world to this new technology and ended the war with two atomic bombs dropped on Hiroshima and Nagasaki. Germany and Japan, however, were close to producing their own atomic weapons. At the end of the war, the USSR captured the secret Japanese military installations in Konan, Korea, and the atomic technology under development there.

Historians know that Soviet ballistic missile research progressed quickly after World War II, and by the late 1950s was in fact dangerously ahead of U.S. efforts. At the time, however, it was not well-known exactly how advanced the Soviet missile threat actually was. The unknown led U.S. officials to imagine the worst. On October 4, 1957, the worst seemed to come true. Without forewarning, the Soviet Union successfully launched Sputnik, the world's first artificial satellite (Fig. 4.2). The satellite, about the size of a basketball and weighing an impressively heavy 183 pounds, orbited the Earth in about ninety-eight minutes in its elliptical path. Before the United States could respond to the call of the International Geophysical Year (IGY) for an artificial satellite to be built, the Soviets had accomplished the stunning feat. Though it wasn't a "shot heard round the world," Sputnik was as effective as one.

This technical achievement captured the world's attention and caught the American public off guard. Fearing the worst, Americans believed that the Soviets' ability to launch satellites could translate into the capability to launch ballistic missiles that could carry nuclear weapons to U.S. soil from halfway around the globe. The public furor over this perceived threat was followed by the U.S. Defense Department's approval of funding for another U.S. satellite project, a backup to the IGY's planned Vanguard satellite. Scientist Wernher von Braun led the Explorer project. A mere month after Sputnik, on November 3, 1957, the Soviets launched a second satellite carrying a much heavier payload and the first animal to enter space, the dog Laika.

On January 31, 1958, less than four months after the first Sputnik launch, the United States launched its first satellite, Explorer I. An exchange of successful and unsuccessful missions between the Soviets and the Americans followed. The American consciousness became captivated,

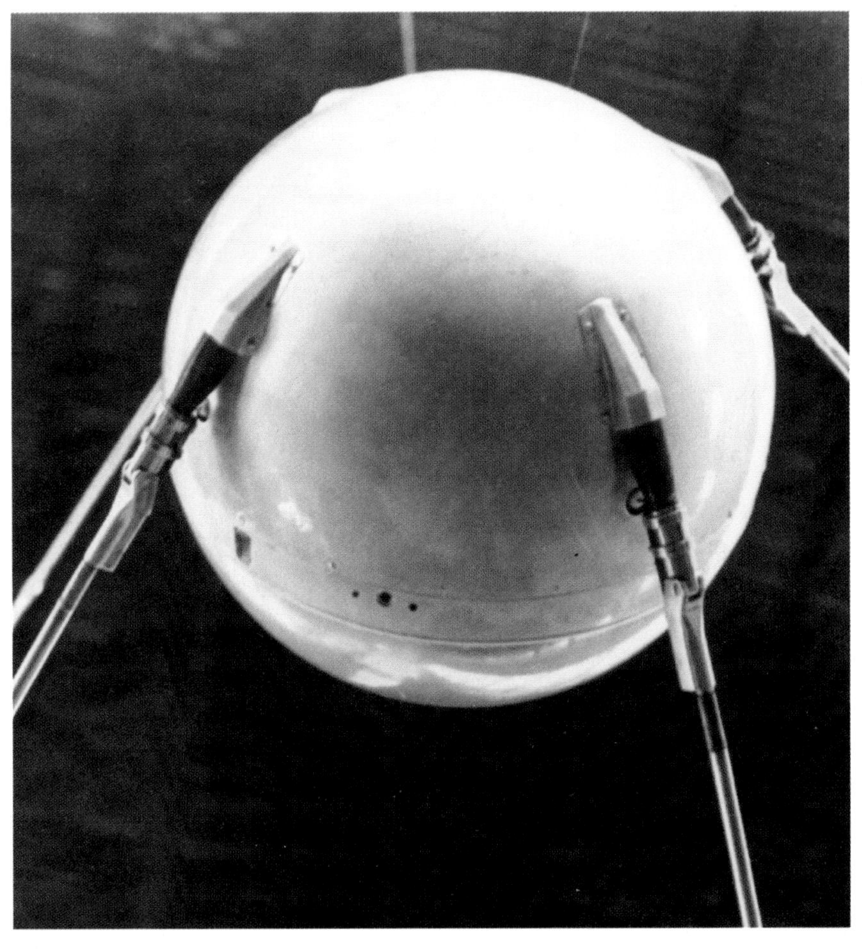

Figure 4.2: *Sputnik, the first artificial satellite, was built and flown by the Soviet Union. On October 4, 1957, it orbited Earth in a little more than an hour and a half. (NASA)*

almost obsessed, with space. In July 1958, Congress passed the National Aeronautics and Space Act (commonly called the Space Act), which created the National Aeronautics and Space Administration (NASA). A new generation of scientists began training as part of NASA.

Although the launch of Explorer I fundamentally signified that the United States could successfully compete with the Soviet Union in the Space Race, the satellite also unlocked a wealth of information about the mysterious near-Earth space environment. This satellite and its successors carried a small scientific payload of instruments that discovered the magnetic radiation belts around the Earth—the Van Allen Belts (named after the principal investigator, James Van Allen). Van Allen knew about cosmic rays (radiation that originates from all over the universe) that reached to the lower atmosphere of Earth. Suspecting that these rays existed in moderate quantities in space, he attempted to use the Explorer satellites to measure them (Fig. 4.3). During one of the Explorer missions, onboard instruments taking data seemed to have failed. The instruments eventually resumed working correctly, and it was discovered that rather than showing nothing, they had been saturated (pegged out) at the unexpectedly high levels of particles. Van Allen found energetic particles that greatly exceeded his estimates; they were extremely high-energy particles, very radioactive, and surprisingly abundant in space. One of his coworkers was to exclaim, "My God, space is radioactive!" The nature of these particles (high energy, high radiation, and large quantity) distinguished them from the milder cosmic rays. They in fact come from the Sun and enter the magnetosphere, becoming trapped in the Van Allen radiation belts (Fig. 4.4).

Figure 4.3: *Three happy scientists (left to right), William Pickering, James Van Allen, and Wernher von Braun, hold aloft the first U.S. satellite, Explorer I. Pickering was the director of the Jet Propulsion Laboratory. Van Allen, the principal scientist for Explorer I, is renowned for his discovery of the highly energized radiation belts around Earth. Von Braun is often given credit for launching the entire U.S. space program. (NASA)*

Manned spaceflights took place during the next phase in space exploration. Orbiting the Earth and traveling farther into space required a new element of space research—the safety of the intrepid astronauts. The newly discovered solar radiation filling space would be dangerous for space travelers. Scientists needed to know more about solar radiation and the methods that could be used to monitor and predict radiation hazards. Aware of the possible danger of space, scientists sent chimpanzees and dogs into space to see whether living creatures could survive in such an alien environment. If all went well, the next step would be to send men into space, starting with short up-and-back missions. Then they would try to orbit the Earth once, followed by longer and longer stints in space. The chronology of human spaceflight followed this small-step process, starting with the first man—a Soviet—to briefly touch space.

The Soviets had beaten the Americans to satellite technology, and they were also the first to successfully launch the first spacecraft carrying a man. The craft, Vostok, took the cosmonaut Yuri Gagarin into space on April 1, 1961. The Soviets had again demonstrated their superiority in space, and this launch sent a frigid jolt through the United States. America was behind, and it raced to catch up. The first

Figure 4.4: *The Van Allen radiation belts (curved blue objects), depicted in cross-section in this diagram, are sometimes thought of as concentric doughnuts around Earth. Electrons that come from the Sun and enter the magnetosphere are trapped in the belts and are revved up to high energies. The high-energy particles make the belts a more dangerous place for satellites and astronauts. (NOAA)*

American in space, Alan Shepard, flew on May 5, 1961, less than a month after Gagarin. It was an extremely proud day for America and for the scientists who had created the technology to make it possible. NASA's historians fondly recount tales of that day. While waiting for liftoff, Shepard had glanced around the cabin and been struck with the thought that "all of this was built by the lowest bidder." Shepard received $14.38 in U.S. Navy flight pay for his fifteen minutes aloft. In the press conference following the flight, the astronaut John Glenn noted that Shepard, whose flight came after the flight of the chimpanzee Ham and before Glenn's own pending orbital flight, represented the "missing link between ape and man."

All joking aside, the United States had serious plans for the future and was determined to do *something* before the Soviets. In a forceful speech, President John F. Kennedy announced in 1962 that the United States would send a manned mission to the Moon by the end of the decade. The roster of spaceflights that followed Shepard's flight shows the back-and-forth race between the Soviets and the Americans to prepare for the ultimate lunar landing: two flights by Americans, one flight by Soviets, two more flights for the Americans, two more for the Soviets, and so forth.

As all Americans know where they were when Kennedy was assassinated, or when the Twin Towers fell, all Americans living in 1969 remember where they were when Neil A. Armstrong walked on the Moon (Fig. 4.5). The sense of triumph was everywhere. People—*Americans*—had conquered space, had learned how to maneuver through it, had survived in its inhospitable environment, and had planted a U.S. flag on the surface of the Moon. Apollo 11 had left Kennedy Space Center (Cape Canaveral had been renamed on December 20, 1963) in July 1969, carrying Neil A. Armstrong, Edwin E. Aldrin, Jr., and Michael Collins. Armstrong and Aldrin rode the Lunar Module to the Moon, while Collins remained in the Command Module orbiting the Moon. The sight of Armstrong walking on the Moon was so unbeliev-

Figure 4.5: *While working at the Canary Islands in the Atlantic, Joe Hirman and his family gathered to listen to the first words from the Moon, spoken by Neil Armstrong. Since there was no TV signal on their island, they looked at maps of the Moon while listening to their shortwave radios tuned to the BBC and VOA. (Joe Hirman)*

able and amazing that there were those who refused to believe the images they saw and swore the landing was a hoax and a conspiracy.

Apollo 11 turned out to be one of a string of lucky Apollo missions that were not hampered by space weather events, which was something of a miracle. Any number of space weather disturbances could have endangered the lives of the astronauts either directly or by damaging equipment. Of course, an astronaut's job is by nature dangerous. Throughout the space program spectacular failed launches and some shocking deaths have occurred, although the casualties have been relatively few for such hazardous work. Scientists and engineers developed launch and satellite vehicles with extreme care to ensure the safety of the crew and the success of the mission. Mishaps did occur that left the control teams and flight crews holding their breath. During such tense times, every relevant expert was consulted, including space weather experts. One example was the drama that unfolded for the Apollo 13 crew.

The Apollo missions had been successful. Both Apollo 11 and Apollo 12 had landed men on the Moon in 1969. Apollo 13 headed to the Moon in April 1970. Its first two and a half days in space went relatively smoothly, despite the ominous number 13 assigned to the mission. Fifty-five hours and fifty-five minutes into the mission (i.e., on the thirteenth day of April) the crew heard a "pretty big bang." An explosion in one of the oxygen tanks had torn off one side of the Service Module (Fig. 4.6). About three hours later, the astronauts found themselves without oxygen stores, water reserves, electric power, or propulsion. The mission controllers in Houston knew that the astronauts were in serious trouble. The goal instantly turned to getting them home as soon as possible—if it was possible. The trip home would take four days—the astronauts had to "slingshot" the ship around the Moon to gain enough power to compensate for the damage to the propulsion system—and all without enough power, water, oxygen, or carbon dioxide scrubbers. Experts on the ground calculated and reappraised each constraint. Use of the oxygen fuel at just the right time and just the right orientation would get the spaceship home; water was preserved by strict rationing, which led to the considerable dehydration of the men; the electric power was partially reengineered and largely shut down to preserve the remaining power, making for a cold cabin; and carbon dioxide was filtered to a level higher than the normal tolerance limit. The crew might well have passed out from CO_2 poisoning had it not been for a bit of science-fair engineering and gray duct tape.

As the ship headed home, everyone grew concerned with the problem of the astronauts passing through Earth's radiation belts in a damaged ship. The Space Disturbances Laboratory responded to questions about how to handle the radiation risk. The radiation levels in the belts

Figure 4.6: *The Service Module on Apollo 13 suffered an explosion that tore off part of the casing, shown here at the left side. The tense return trip made for a heroic tale that ended with sighs of relief for all. (NASA)*

around Earth were not elevated, thanks to low solar activity. In the undamaged part of the ship, the Command Module, the crew would have good shielding. Because they were coming straight home, they would not be in the radiation belts for long. It was decided that the radiation belts would not be one of the many dicey conditions and constraints suffered by the astronauts.

The crew landed safely, owing to some amazing engineering and some luck. Analysts later learned that human errors and technical anomalies had caused the explosion. This close-call mission drove home the point that space travel is risky. Although there would be more manned flights into space, NASA officials began to think that little more could be learned from visiting the Moon again. Unmanned observational satellites, capable of constant patrol, could study stars, planets, the Sun, or the Earth without needing food, water, or oxygen. Self-contained satellites would be the greatest boon for space research.

By the late 1960s, American interest in space had begun to turn toward research. The Soviet threat had diminished slightly since, after all, the United States had been first to reach the Moon, the ultimate prize at the time. The successes and failures of launches had taught scientists a great deal and would help them in the next phase of space exploration—satellite research. Satellites could be used to gather data on the space environment and, as it would turn out, those data would be used to protect commercial satellites from the same hazardous environment. At first satellites looked primarily at terrestrial weather for military and civilian purposes. The need for real-time weather forecasting led to geostationary satellites, which in turn allowed for better observations of the Sun. Space-based solar observatories followed, and as technology improved, so did the satellites and the instruments they carried.

Weather observation from space was a top priority for both civilian and military users. The ability to actually see the weather patterns over the entire Earth was a dream come true for meteorologists. NASA started satellite observations for civilian use in 1960 with the launch of the Television InfraRed Observation Satellite (TIROS-1), the first in a series of weather satellites under that name. The satellites were polar-orbiting, meaning that they crossed over both the north and south poles in their ninety-minute orbiting cycle. The Environmental Science Service Agency (ESSA, later to become the National Oceanic and Atmospheric Administration [NOAA]) provided space weather monitors for the TIROS satellites. When TIROS-9 went into orbit in 1965, space weather observers began to receive complete daily coverage of the entire Earth. By the time TIROS-10 (also called ESSA II) was launched, ESSA engineers operated the satellites and assumed all responsibility for the satellite.

In 1979 a new generation of TIROS, the TIROS-N/NOAA Program, was launched. NOAA, not NASA, built and currently runs these polar-orbiting satellites. In the near future, these satellites will be replaced by the National Polar-Orbiting Operational Environmental Satellite System (NPOESS), to be used to monitor global environmental conditions and collect and disseminate data related to weather, atmosphere, oceans, land, and the near-space environment. NPOESS will combine the existing polar satellite systems from the Department of Commerce (DoC) and Department of Defense (DoD) and will result in a more cost-effective and better-integrated system. The total constellation should be operational in 2013.

Research Instruments in Space

Geostationary Operational Environmental Satellite Series

By the time NASA launched TIROS-10 in 1965, the agency already had in mind a future geostationary satellite that would provide better images and weather data. The geostationary, or geosynchronous, orbit is a unique and valuable orbit. As the name suggests, objects in this orbit remain stationary relative to Earth—that is, move at the same rate as the Earth. Roughly 22,300 miles from the surface, the satellite "falls" toward Earth at the same speed as Earth rotates on its axis. Following Kepler's Laws, the satellite stays in the same place relative to a point on the Earth over the equator. If you were to take a time-lapse exposure of the sky, holding the camera steady, your picture would show all the stars as streaks (because of the Earth's rotation), but the geostationary satellite would show as a single clear point (Fig. 4.7).

Figure 4.7: *A time-lapse exposure of the geostationary sky reveals the satellites as dots that do not move relative to us on Earth. The stars that move through the night sky as the Earth rotates show up as streaks. (Paul Maley)*

Tracking a geostationary satellite requires a *fixed* antenna rather than an antenna that must pivot to track the satellite as it moves across the sky. Satellite TV works because a single fixed dish can constantly receive programming day and night from a geostationary satellite. Meteorologists can watch the weather develop over a specific location because the satellite always takes pictures of the same spot. Space weather forecasters like the geostationary orbit because it is far away from the Earth and usually not obscured from the Sun.

The Geostationary Operational Environmental Satellite (GOES) series of satellites began the next advancement in weather-watching. At least two satellites were always in orbit at the same time so that if one failed, the stream of space weather data would not run dry. They were continually operational, downlinked data in real time, and carried crucial space environment instruments (Fig. 4.8). With the launch of GOES in 1971, real-time weather information flowed to NOAA, and a small part of the transmission carried space environment sensor data directly to Boulder. This vitally important real-time aspect made space weather forecasting possible because scientists could respond before an event occurred rather than at or after the time of the event. (Forecasting and the implications for customers will be discussed more in Chapters 6 and 7.)

Figure 4.8: *The NOAA GOES satellites are the operational weather satellite series for the country and a vital part of National Weather Service observations. This rendering shows the newest design series built by Boeing and launched in 2005; the previous series satellites were launched between 1994 and 2002 and will be phased out by 2008. (NASA Jet Propulsion Laboratory)*

One might think that actually *seeing* the Sun would yield the most important information about the space environment, but that notion fails to take into account the most valuable data scientists have. Although thought of as a weather-watching satellite, GOES has been the workhorse in collecting solar measurements of solar protons, electrons, and x-rays and to this day produces the most widely used data and longest-running data records in the space weather community. The

Space Environment Center receives all types of data within seconds of GOES collecting it. Although the computer systems that collected and organized the data at first were hard to manage, requiring difficult command lines and many reboots, the usefulness of the data was state-of-the-art. The consistent GOES space weather data reaching back to 1972 are the gold standard for research and forecast modeling and are comparable only to the sunspot data records collected by ground-based telescopes since the 1700s.

Skylab

The first space-based solar observatory, Skylab, launched in May 1973, carried more instruments than any other satellite to date and required a human presence. Skylab was a space station that operated only for about two-thirds of a year. Because the instruments were not automated, as in most satellites, the station floated "blindly" in space when no people were there to operate it. The station carried the Apollo Telescope Mount (ATM), which held all of the instruments. The ATM was as large as a ground-based observatory and loaded with instruments that could capture a great range of solar radiation through Earth's atmosphere (most not visible), a great range of space (from narrow to panoramic features), and several different wavelengths (for looking at the different layers of the Sun). Because a crew manned the space station for part of the time, NASA scientists were concerned about the radiation hazard. NASA relied on space weather alerts and forecasts for many aspects of this program. Its Solar Radiation Analysis Group (SRAG) worked especially closely with the Space Environment Services Center (SESC) to oversee the safety of the flight crew.

The astronauts assigned to work on Skylab needed training in solar observing and the array of instruments they would use (Fig. 4.9). They traveled to Boulder to receive that training from SESC as well as receiving briefings before launches and during the missions. Scientists interested in doing research with Skylab needed help from SESC staff to pinpoint the areas of the Sun that were likely to see flares or coronal holes, or whatever the scientists wanted to study. Daily meetings occurred between SESC forecasters and research scientists; the researchers posed the big questions, and SESC helped by getting the specific images or data to answer those questions.

Skylab images, carefully selected by scientists on the ground and taken by astronauts in space, profoundly improved our understanding of the workings of the Sun. Each instrument told scientists something different about the Sun (Fig. 4.10). The x-ray imager showed dramatic pictures of the solar atmosphere. Spectrographs told of the chemical composition of the atmosphere. The coronagraph (similar to the one

Figure 4.9: *Staff from NASA and NOAA at the Houston Space Center during the Skylab missions not only supported the astronauts in the space environment but also helped the scientists observe the Sun in novel and revolutionary ways. (NASA)*

Walt Roberts had operated at the Climax mine) showed the outermost layer of the Sun and revealed some striking information about the Sun. The corona, which had been viewed naturally during solar eclipses for millennia, appeared hazy white around the edges of the Moon in a total eclipse. During those rare eclipses, observers could catch glimpses of wispy material streaming away from the Sun.

Skylab had an excellent location for viewing the corona. Its coronagraph saw the sudden bursts of accelerated matter coming out from the Sun, seemingly through magnetic loops that would break open. These bursts of matter, called coronal mass ejections (CMEs), shoot megatons of ionized gas (plasma) into space and potentially disrupt satellite signals and power grids on Earth. Scientists working on Skylab had hoped to see such ejections but considered them as rare as unicorns. When the

Figure 4.10: *Key characteristics of temperature, density, and altitude in the Sun's atmosphere. The Sun's atmosphere is dramatically different from the Earth's. The highest density on the scale is still only as dense as Earth's atmosphere at 90 km (56 miles) above sea level. The melting point of silver is near the coolest part of the Sun's temperature range. (NASA)*

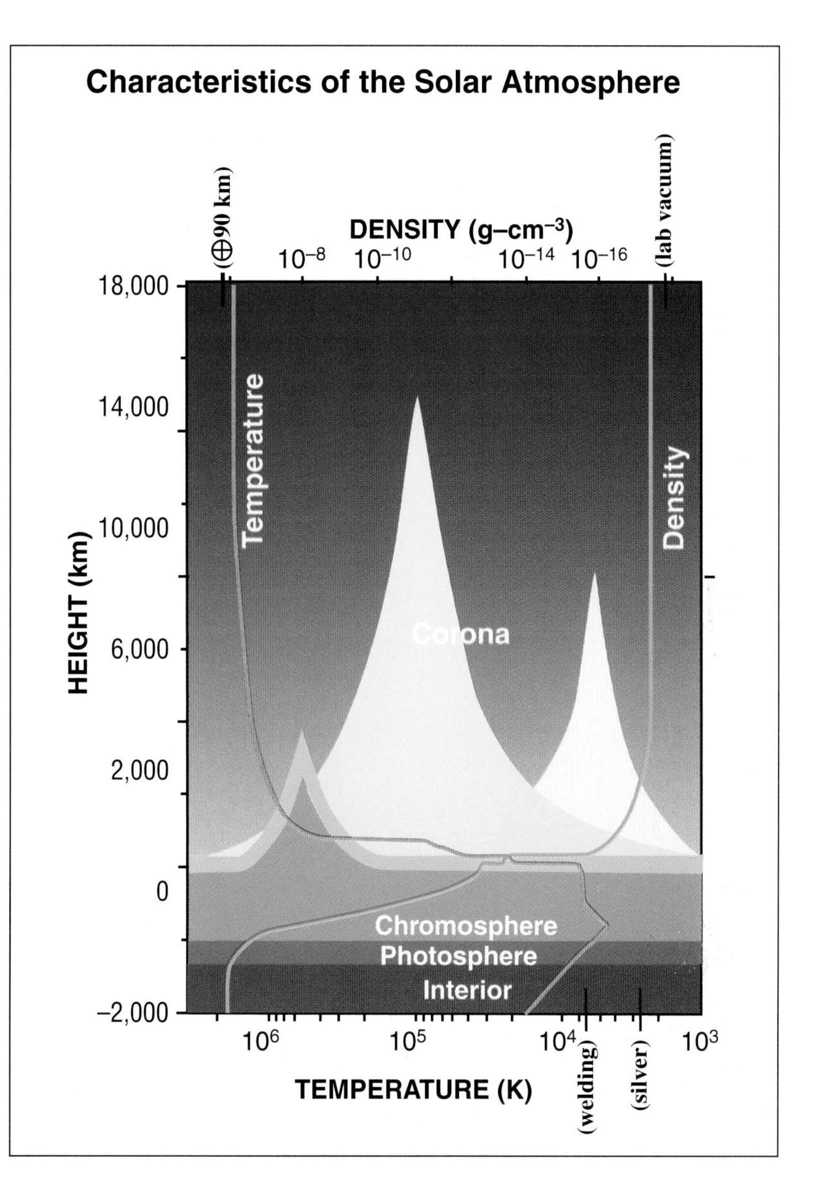

pictures from the coronagraph came back to the scientists, they whooped and hollered when they saw that there were hundreds of CMEs. The scientists learned from Skylab that the CMEs sometimes occur with solar flares, but not all the time. Solar flares originate in the active regions of the Sun, and the instruments from Skylab revealed the complexity of the sunspots and active regions. In and around sunspots, scientists saw an intricate magnetic structure that was constantly changing. Magnetograms (taken through a telescope that showed magnetic fields) clearly showed patches of very active and disturbed regions likely to emit flares, and consequently more likely to emit CMEs.

Another major Skylab discovery was the resolution of the mysterious magnetic regions (M-regions) thought to be the sources of recurrent geomagnetic storms. By combining the x-ray wavelength images from Skylab with newly available high-resolution magnetograms taken from the ground, scientists uncovered the mystery of these features and named them coronal holes (Fig. 4.11). A coronal hole was a massive low-density region, covering perhaps a quarter of the surface of the solar disk, that tended to form around the north and south poles during Solar Minimum. Perhaps most exciting for scientists was the discovery that coronal holes were the long-sought source of the high-speed solar wind. The solar wind, like terrestrial wind, normally blows in a constant high-density, low-speed stream (more or less a background flow), but it also can blow in low-density, high-speed streams during solar events (the "enhanced" solar wind). Because of Skylab, scientists figured out that when a coronal hole aligns to face the Earth, and the intensity of the solar wind increases, negative effects on Earth systems due to geomagnetic storms are likely.

After the third astronaut team visit to Skylab, NASA allowed the space station to fall silent. Although there were no plans to use it again, NASA still had the responsibility to track the satellite and either bring it down safely or push it out into space. Until they dealt with it, Skylab would

Figure 4.11: *The dark area, dubbed a coronal hole, was one of the startling and most important discoveries of the Skylab mission. Visible only through x-ray images, these cooler and weaker magnetic areas were found to be the source of the solar wind. (NASA)*

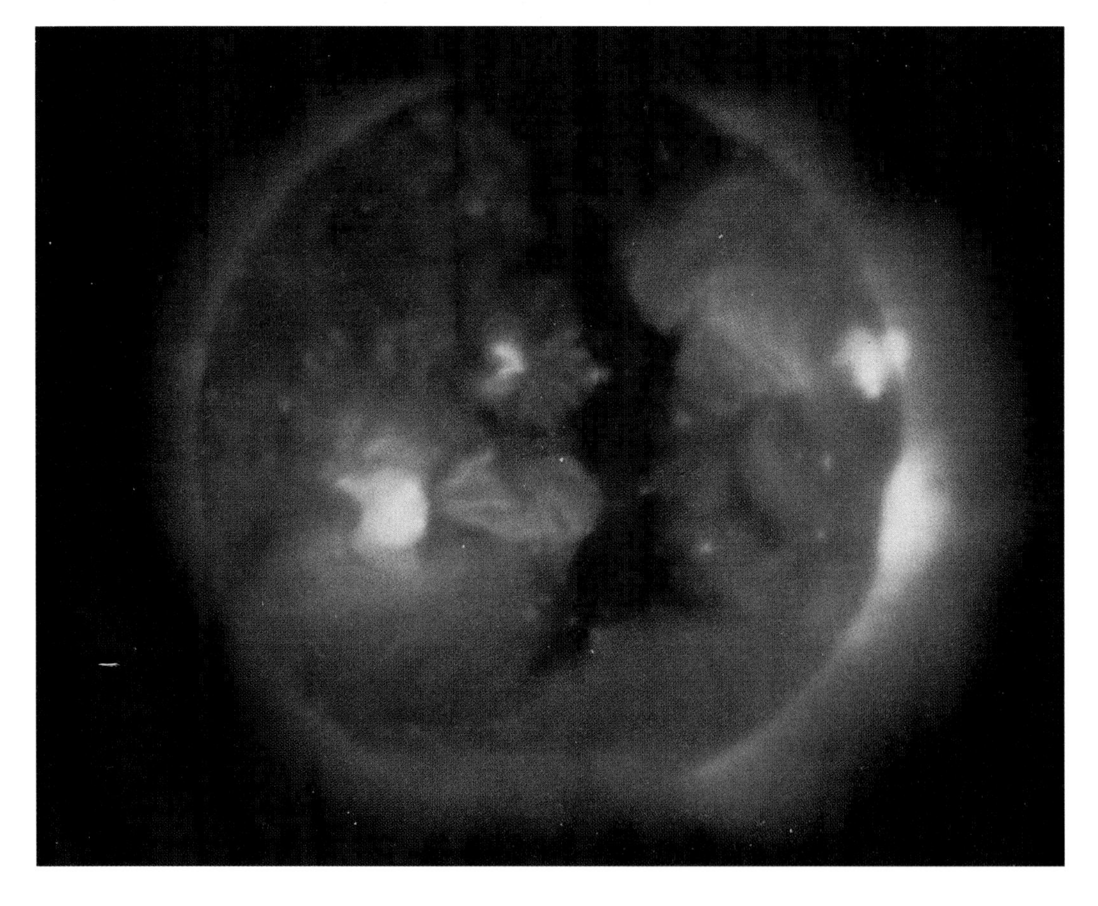

remain in its normal orbit—at least that was the plan. NASA forecasters had predicted that the next sunspot cycle would be a low one, contrary to NOAA's prediction. High solar activity (usually associated with high sunspot numbers) can disrupt the orbit of satellites. NASA underestimated the size of the sunspot cycle and so was surprised when the drag on Skylab due to space weather effects was stronger than it had expected. Disturbed from its orbit, Skylab fell to Earth in burned pieces seven years later, in 1980. After that, NASA left the predicting of the solar cycle to NOAA. The discoveries made by Skylab scientists not only greatly expanded the reservoir of knowledge about the Sun but also gave scientists a glimpse into the future of space weather forecasting and observational systems that was still two decades away.

Solar and Heliospheric Observatory

Thanks to the first-of-its-kind Skylab observatory, NASA scientists had a clear idea of what they wanted in terms of the perfect research satellite. In 1995 NASA and the European Space Agency (ESA) jointly built the Solar and Heliospheric Observatory (SOHO). The two agencies operate it to this day. Like Skylab before it, SOHO had a wealth of advanced instruments in its payload. Unlike Skylab, SOHO did not need astronauts to operate it, and its mission would not be as short; it was a research satellite, not expected to support operational forecasting. The staff of the Space Environment Center (SEC) worked with SOHO scientists to set up the transmission of data all the time in real time, which would serve the operations sector. (A second backup satellite, which would ensure data, would be required to make the satellite technically "operational.") To make the best use of this valuable satellite, countries around the globe agreed to track it and record data. As the satellite passed out of range of one station, the next station along the path of orbit would begin collecting data, ultimately giving forecasters a continuous stream of data.

For SOHO to get the best look at the Sun, NASA chose the L1 orbit. Like the geosynchronous orbit, the L1 orbit is unique. In 1772 the French mathematician Joseph Louis Lagrange deduced the orbits of a three-body system in which all the bodies retain a stable orbit vis-à-vis each other. In stable orbits, gravitational forces and the orbital motion of the bodies balance each other. Lagrange named five distinct orbits that had special properties. An object at L1, L2, or L3 is semistable, like a ball sitting on top of a hill—a little nudge and the ball will roll away. A spacecraft at one of these points has to frequently fire its rockets a tiny amount to stay "on the hilltop." SOHO sits in orbit about a million miles Sunward from Earth in the L1 position. Because it stays between the Sun and the Earth at all times (which is the main reason that scientists chose this orbit), the observatory can stare at the Sun without the Earth or Moon eclipsing it. Incidentally, an object at L4 or L5 is truly stable, like a ball in a bowl; when gently pushed away, it orbits at the Lagrange point without drifting or needing frequent rocket firings (Fig. 4.12).

Following a final checkout in low-Earth orbit, SOHO was moved into orbit at the Lagrangian L1 point of the Sun-Earth system. As L1 lies very far from the Earth, most satellites destined to orbit at that point have to make quite a trip to get out there. Satellites often slingshot around the Moon to gain "free" momentum for the long trip. Some satellites have even gone twice around the Moon before traveling on to reach an orbit that is, at times, 250 Earth radii away. NASA and ESA declared the SOHO mission fully functional in April 1996. SOHO carries a suite of instruments more advanced than any other solar observer. The images taken in several different wavelengths are false-color coded to make them instantly recognizable as specific layers of the Sun. SOHO has two perspectives of the corona, one showing the corona out to about four solar radii, the other to six solar radii.

On January 6–7, 1997, SOHO scientists observed a large and impressive head-on view of a CME. Don Michels, a NASA scientist, indicated that he had little doubt that it was a "halo CME," meaning the blast of material was aimed directly at the Earth. Seeing the blast delighted the scientific community because of the dramatic certainty that trouble was headed straight for Earth. Observers had seen ejections from the Sun

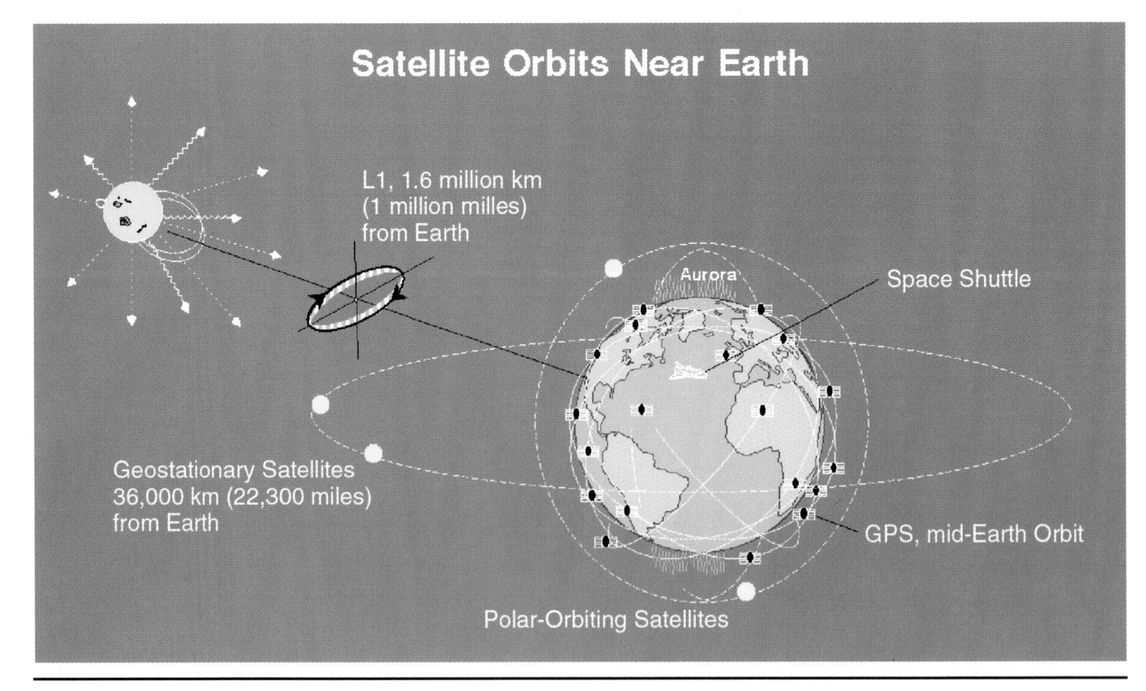

Figure 4.12: *Commercial and observing satellites orbit at many different altitudes. (NOAA)*

before but could not be sure how much of the material would reach the Earth or what the impact would be. The impact of a halo CME was a bit more predictable. This event illustrated how valuable the coronagraph was as a tool.

On April 7, 1997, the Large Angle and Spectrometric Coronagraph (LASCO) imager (one of SOHO's coronagraph instruments) captured another halo CME. Excitement ran high among SOHO scientists, and their glee led to some sensationalist press coverage (Fig. 4.13). Dan Rather's script for the April 10, 1997, CBS Evening News read:

> **Rather:** The first shockwave from a massive solar belch is zapping planet Earth tonight. The sun-watching satellite SOHO caught the scene, as the solar storm was ejected and propelled toward Earth at 2 *million* miles an hour. There's no danger to humans on the ground, but it could fry some expensive communications satellites in orbit. CBS News Correspondent John Blackstone is on the Earth alert, and watching the skies.
>
> **Blackstone:** Unseen by most of us on Earth, there's been a raging storm on the surface of the Sun this week. Satellite photographs show the massive eruption, says NASA astronomer Steve Maran: "We had a huge coronal mass ejection—a great outburst of matter and magnetic fields from the sun." Some scientists call it a solar belch, sending a massive cloud of electrically charged solar particles right this way, traveling at about two million miles an hour. Says Maran: "It's like a missile that's heading right for you—when it starts looking bigger you know to duck. The Earth can't duck." When the cloud hits tonight, telephone and television service could be disrupted, satellites could be damaged but the Earth's magnetic field should shield people from the radiation. The north and south poles will act like lightning rods taking a surge of hundreds of billions of watts of electric power. If even a little of that surge hits the electrical grid, the lights could go out. Ben Damask of the Electric Power Research Institute says: "This one seems to be a big disturbance but it's hard to predict whether or not this storm is going to have an effect on the power grid." ... Says Maran, "Scientists are really excited about this because we've never had 'til recently the capability to track one of these solar eruptions as you might say from cradle to grave."

This somewhat dramatic press release reflected the excitement of the scientists and let loose the imagination of the public. Despite the clear statement by Dan Rather that there was "no danger to humans on the ground," SEC received calls from hospitals, police departments, emergency managers, and citizens—all wanting to know how to prepare for this supposedly Armageddon-like event. The resulting solar storm, as predicted by SEC, was not particularly big and failed to concern the forecasters; in this case, the story outran the facts. The unprecedented

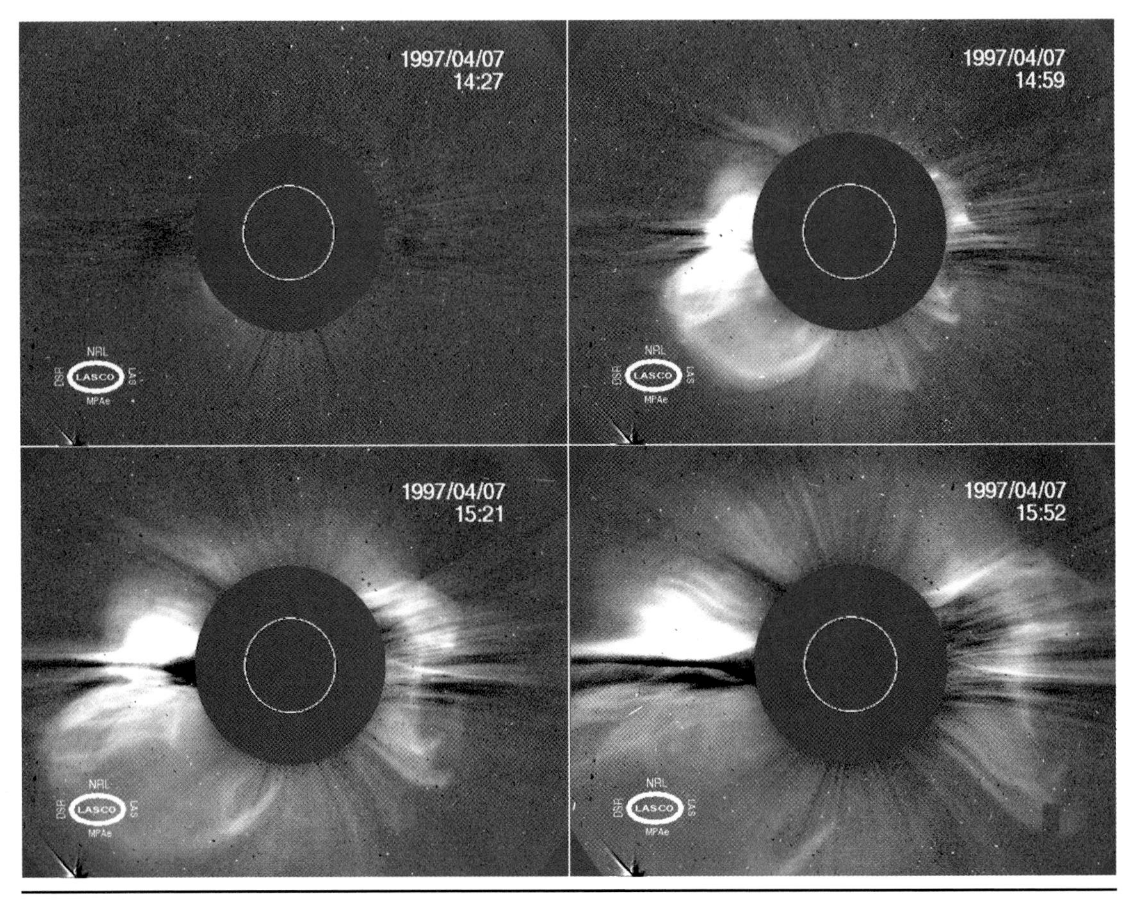

Figure 4.13: *The CME on April 7, 1997, excited scientists and the press. The sequence of images was taken by the LASCO coronagraph on SOHO. This image gave scientists a clearer representation of the dynamics of this "halo event." (NASA/ESA)*

and unnecessary panic brought on by dialogues between NASA scientists and the media led to a high-level meeting among twenty people from NASA and SEC. They agreed that future uproar should be avoided. NASA would not predict space weather or intimate any possible impacts on the public, and NOAA would remain the nation's purveyor of space weather forecasts and warnings.

SOHO became the darling of forecasters and scientists alike. Its value to the community was huge. Then, on June 25, 1998, it was suddenly lost. Usually in nearly constant communication with the satellite, scientists could not make contact with it and therefore did not know where it was. The incident happened during a period of planned calibrations, maneuvers, and spacecraft reconfigurations. Speculations about the cause ranged from a misbehaving thruster that could have moved the satellite to an explosion or other unseen accident. A slight rotation could have allowed the satellite to turn away from the Sun. Without solar energy falling on the solar panels, it would quickly run out of power—and out of power means out of luck. The satellite would be helpless to right itself, talk to mission control, or receive instructions from Earth. Furthermore, without solar-powered temperature controls, its instruments would face extreme temperatures; the side facing the Sun would bake while the other side would freeze.

Alan Kiplinger of the University of Colorado and his colleagues, along with an international group of volunteer astronomers, went looking for SOHO, using their ground radiotelescopes to search the place where it had last been seen. The station in Arecibo, Puerto Rico, aimed a radar signal at where SOHO should have been and was delighted when a station in California began to receive a faint echo back. They had found it! The satellite was where it should be but rotated in such a way as to leave it powerless. Clever mission scientists calculated when the passive satellite would drift and rotate back into an orientation where light would charge the panels a little. In August, with the panels minimally charged as expected, the scientists sent a message to the satellite, which could now receive instruction. They managed to orient the satellite so that it could recharge, power up, slowly thaw out, and begin functioning again. It had taken about a month to regain this valuable asset.

For his work in the SOHO recovery mission, Alan Kiplinger received a letter of thanks from the ESA that read:

> Paris, 3 August 1998
>
> Dear Dr. Kiplinger,
>
> With this letter I would like to express my appreciation and gratitude for your special efforts in the radar detection of the SOHO spacecraft on July 23, using a beam from the NAIC radio telescope in Arecibo. This would not have been possible without your initiative and encouragement.
>
> As you know, from these observations we could confirm the position in space of SOHO and obtain an estimate of its spin rate. This information has been extremely useful, and given us reason for optimism for future recovery.
>
> Please accept the gratitude of myself and of the international scientific community for your outstanding efforts.
>
> Yours sincerely,
>
> R. M. Bonnet (Director, Scientific Programme)

According to the SOHO Mission Interruption Joint NASA/ESA Investigation Board, a routine instruction to the ship "contributed to the initiation of a calamitous sequence of events. [The procedure] had been developed to simply reconfigure the three roll gyroscopes after calibration; however, the procedure had also been modified to provide options to perform gyro spin down." In trying to recalibrate the gyros, scientists had accidentally shut the gyros down—and loss of the gyros is indeed calamitous. The board concluded that no anomalies had occurred on board SOHO but that a number of ground errors had led to the major loss of attitude (rotation) experienced by the spacecraft. Although human error had caused the loss of the satellite, engineers and scientists had also, incredibly, saved it; very few satellites lost in this way are recovered. The return of SOHO to an operational state brought great relief, especially to SEC forecasters. The crisis brought to light just how fragile the daily data link to satellites is, and the vital importance of a backup satellite in an operational suite (something SOHO lacked). None of the agencies responsible for SOHO had any intention of duplicating the valuable satellite, but its proven value made it clear that the critically important instruments on board needed to be replicated on other spacecraft to keep data flowing.

SOHO remains in orbit as of 2006, continuing to deliver valuable images. NASA maintains the latest SOHO images, usually only hours old, on its website (http://sohowww.nascom.nasa.gov/). The site also provides movies that are eye-opening and beautiful. Like other satellites, SOHO will inevitably die, but everyone hopes it has a long life ahead. Looking to the future, NASA is working on follow-up missions that will observe some of what SOHO observed. For now, scientists count their blessings (Fig. 4.14).

Solar X-ray Imager

Skylab, with its great array of instruments and possibilities, whetted the appetite of the NOAA scientists and forecasters who wanted an instrument that could provide a ready source of frequent, fresh solar x-ray images. It would take years to turn the dream of the solar x-ray imager (SXI) into reality. The whole process, from developing the instrument to finding "a ride" on a "bus"—a satellite to carry it into space—to operating the orbiting instrument, and finally analyzing the images and data, took the participation of NOAA, NASA, and eventually the U.S. Air Force (USAF).

In 1966 Dick Grubb (SEC) and Bob Kreplin at the Naval Research Laboratory (NRL) wrote an initial proposal to NASA to build a prototype for an operational solar monitor in space, beginning a project that would take twenty-five years to complete. Several years later, on the lawn of the University of Colorado stadium in the midst of a Simon and Garfunkel concert, Bill Wagner and Gary Heckman chatted about the intense need for the instrument. They set in motion the next phase of the campaign. Wagner (a NASA researcher), Heckman (a NOAA SEC forecaster), and Pat Mulligan (the programmatic overseer in the NOAA satellite center in Washington, D.C.) made up a team that would find

Figure 4.14: *The Transition Region and Coronal Explorer (TRACE) satellite, developed by the Stanford-Lockheed Institute for Space Research, has a unique view of the Sun's surface and magnetic fields. TRACE scientists coordinate observations with SOHO by obtaining simultaneous measurements on both the fine-scale magnetic fields and the associated large plasma structures on the Sun. (NASA)*

funding and support for the new instrument. Together they created reports and displays and "hit the road." They held on the order of sixty meetings at NOAA, USAF, and NASA, explaining the improvements this imager would bring to space weather forecasting. The Air Force saw the great advantage in monitoring x-ray emissions so as to improve the forecasting of ionospheric disturbances. It had great interest in the anticipated data from SXI—and it had money. Michael Jamilkowski, the procurement officer within the USAF, recognized the value of SXI in an interview filmed for an SEC outreach video: "We communicate through that layer [the ionosphere] ... everything we do is affected by that layer and it's very important. The Air Force was ready to support the mission. NASA was too. The Department of Defense came up with some funding, NASA actually came up with the people and the manpower to build it, NOAA had the spacecraft [GOES] and the funding to launch it. So with all three agencies together it was a very, very good job for the taxpayer." As the project approached the launch date, Heckman, now the senior forecaster at SEC, said in 2002 (also filmed for the outreach video), "I can't even anticipate what kind of improvements we're going to have [in ten or fifteen years], but I'm sure it's going to be revolutionary; it's going to be exciting."

SXI has truly proved to be exciting. The small, fifty-pound instrument hitchhiked a ride on the NOAA GOES-12, launched on July 23, 2001. When the spacecraft was in place, the instruments were turned on. Even before the operators had reason to believe that they had the spacecraft pointed at the Sun, SXI had captured data from a brilliant flare on the Sun. This success marked the culmination of significant career efforts, and all involved celebrated. Once all systems were checked out, NOAA put the spacecraft, with SXI on board, "in storage," ready to be turned on when an older GOES failed. "Having this satellite to back up the GOES system is very important," said Steve Kirkner, NOAA's GOES program manager, in comments published on the NOAA website. "If one of the older GOES satellites fails, GOES-12 can be pressed into service without delay. With GOES-12 stored in orbit, we will be able to receive data within two days of activation."

The case for having a backup satellite was clear, but it brought some short-term anguish. The triumph of having succeeded in getting an SXI into orbit seemed ephemeral as it floated, silent and blind, in space. With operational suites, scientists find themselves in a terrible bind: they must have backup equipment in case the original satellite fails, as an uninterrupted flow of data is vital; however, any new instruments placed in space will begin to "weather" and age whether turned on or not. In the case of the SXI instrument, the latter concern won out, and NOAA turned on GOES-12 again in December 2001. Although no pressing need existed to replace the already working GOES satellites, scientists figured that SXI might as well keep its eyes open. SXI began to collect real-time data, sending down images every few minutes. It suffered several glitches and sustained damage over the years, yet, with diminished abilities, SXI continues to bring smiles to the faces of the forecasters and scientists who use its data.

Other Satellites

Four research satellites sent up to monitor the solar wind, not collect images of the Sun like the others mentioned here, laid the foundation for quick, accurate forecasting of geomagnetic storms. Previously geomagnetic storms could be predicted only three to four days before their arrival at Earth, and nothing could be known in advance regarding how severe the storms would be. By measuring the solar wind, which carries geomagnetic storms, forecasters could give the public more accurate, useful information. Pioneer-5 provided one of the first measures of the solar wind in 1960. The satellite orbited the Sun between Earth and Venus and, as one of the earliest satellites, was the first to communicate with Earth from such an immense distance—a record 36.3 million km (22.5 million miles). The Pioneer-5 satellite and successive Pioneer crafts recorded data on cosmic rays and the solar wind. Pioneer-5 also mapped the interplanetary magnetic field for the first time. During a time when most of the missions into space lasted no longer than a few days, Pioneer-5 functioned for a record 106 days. Some of the instruments on the Pioneer probe series still work today, forty years later—and thirty-nine and a half years longer than their expected life!

The International Sun Earth Explorer (ISEE) ran a string of satellites, the third of which (ISEE-3) went into orbit on August 12, 1978. Scientists designed ISEE-3 to monitor and study the interplanetary medium and the magnetosphere, which of course meant that it had to be continuously outside Earth's magnetosphere. ISEE-3 was the first spacecraft to use the L1 orbit, as that would place it far outside the magnetosphere, 235 Earth radii away on the Sunward side of the Earth. From there, it continuously monitored the energy-transfer processes between the solar wind and the Earth's magnetosphere. ISEE-3 transmitted a steady stream of data back to Earth, which required someone to be "listening" at all times. This system had real promise for the forecasters who needed a sentinel "upstream" from Earth to issue warnings of impending storms for operators of power plants and pipelines. However, ISEE-3 finished its mission of monitoring the solar wind in 1982, and scientists moved it to study the comet Giacobini-Zinner. In 1991 NASA surprisingly extended the already-old satellite's mission to return to gathering data on coronal mass ejections and cosmic rays.

The data on the solar wind coming back from Pioneer and ISEE-3 thrilled scientists, and they wanted more. USAF and NASA funded the Wind satellite and launched it to L1 on November 1, 1994. Its research mission, like ISEE-3's original assignment, was to gather data about the magnetosphere and the region of space upstream from Earth. The Wind satellite provided information about the speed of the solar wind and the energy of the particles carried by it. Although the solar wind comes to Earth at 1.6 million kilometers (one million miles) per hour, the measurements taken at L1 travel to Earth at the speed of light, giving forecasters an edge in predicting what is coming and when. Wind did a fair job of delivering data, but no one continually tracked it, so the scientists did not always receive real-time data.

NASA's Advanced Composition Explorer (ACE), launched in 1997, orbited at L1 and monitored several aspects of the solar wind. However, NASA had no plans for collecting data in real time from this satellite. ACE's Solar Wind Monitor (later renamed the Real-Time Solar Wind monitor) recorded data continuously, but as with other satellites, the data were not real-time unless the satellite was tracked by antennas around the world. Ron Zwickl and a team of SEC staff took on the job of coordinating international partners to do the tracking. By January 1998, the myriad details of finding equipment, responsible personnel, and compatible transmission systems had been diligently worked out (Fig. 4.15).

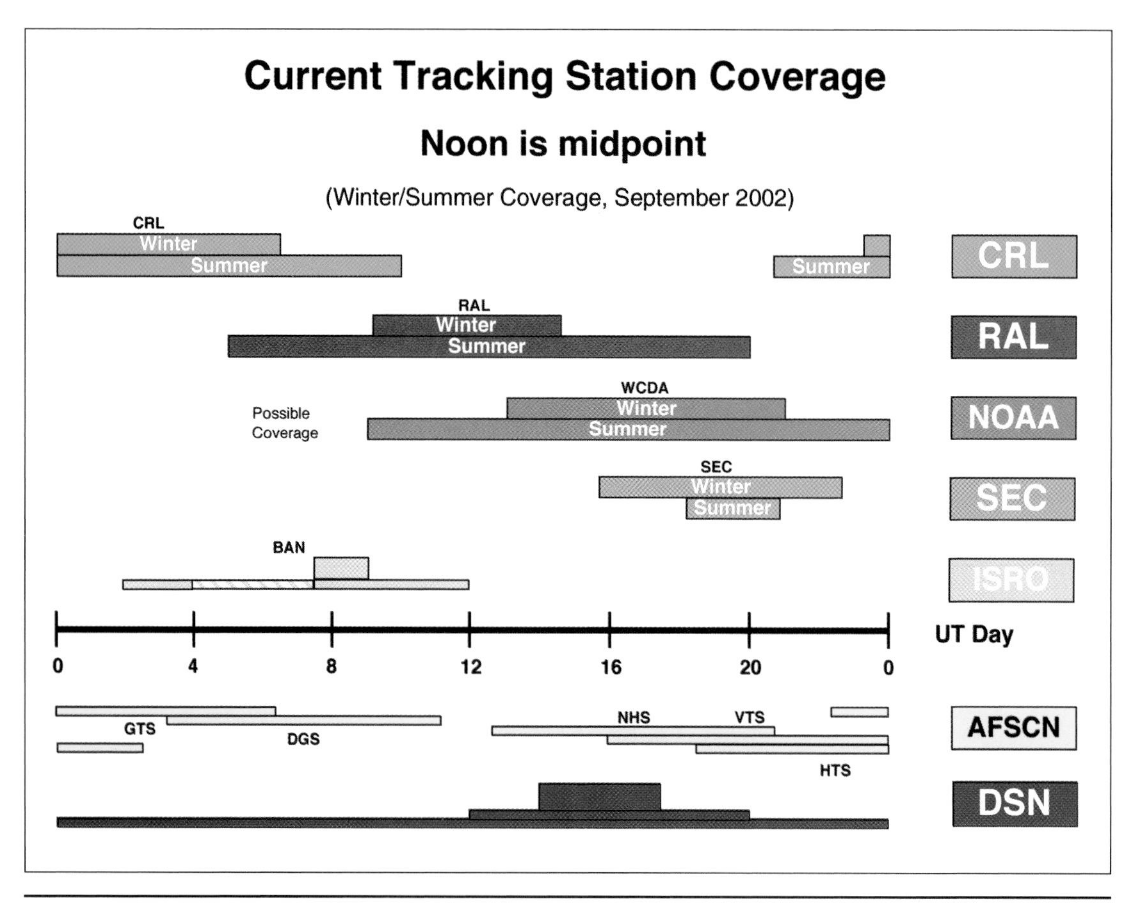

Figure 4.15: *Tracking ACE is made more complex by the varying angle of the Earth throughout the different seasons. Worldwide tracking of data from ACE depends on Japan, the United Kingdom, NOAA, India, the U.S. Air Force, and NASA. (NOAA)*

ACE was the first instrument that let scientists accurately warn customers of impending geomagnetic storms regarding how severe the storms would be, how soon they might arrive, and how long they could last. Depending on the speed of the solar wind, the advance warning of a typical storm is only thirty to ninety minutes, but that lead time is substantially more than it was before. ACE's orbit has been changed slightly so as to reduce its fuel consumption and keep it limping along through its declining years. Replacing it is vital, so SEC is working to get a new operational satellite to fly solar wind monitors as soon as possible. Forecasting of geomagnetic storms will be severely compromised with any loss of solar wind data.

Satellites, born out of the Cold War, greatly matured our view of space weather. They became unblinking eyes that looked at the Sun, the Earth, and the busy space in between, with long patrols that could not be undertaken by manned missions. Ultimately, satellites pushed space weather forecasting from a dark art to a serious technical-support tool. Of course, it took more than just satellites to do that. It took a coherent, committed "army" of observers, forecasters, technicians, and engineers using advanced equipment and techniques to build a space weather service that would serve the world.

5 Building a Service on Observations

The "space environment" until recently has been thought of only as the space between the Sun and Earth. The Sun spews out energy and plasma in all directions, but because much of that energy and plasma will never come near Earth, we have, in our self-centered way, disregarded a huge percentage of the space that can be affected by the Sun. Yet as we expand our sphere of interest, say in a manned flight to Mars, we must address the full volume of space around the Sun. The space environment is huge! Consider the volume that contains the variability of weather on Earth: the surface and atmosphere is about 14.5 km (9 miles) thick. In that space we have monitoring devices and stations all around the Earth and a few satellites with data-rich collecting instruments; that amounts to so much data that weather modeling now requires supercomputers to do the calculations. Now consider the volume of space starting at the Sun (into which you could fit one thousand Earths) and reaching to the Earth—150 million km (93 million miles). For this volume (not even the entirety of the space environment) we have a few telescopes and a few satellites. The space environment is several orders of magnitude larger than Earth's weather environment, but the task of monitoring and predicting the environment is essentially the same.

So whereas terrestrial weather predictions rest on a large amount of data for a small amount of volume, space weather predictions rest on a tiny amount of data for a huge amount of space. The increase in the number of satellites during the last half-century did increase the amount of data scientists had to work with; however, without worldwide tracking of the satellites, the data were not always useful for forecasting. Advanced communication was required to keep the Sun under twenty-four-hour-per-day observation. The logistics of creating a worldwide network of observatories were understandably complicated, involving money, staffing, and management: initial capital to built the observatories and equip them, qualified staff to run the observatories, a maintenance budget to keep the equipment up-to-date and functioning, and tight coordination among the observatories.

Despite the difficulties in establishing a worldwide network of observatories, the space weather community that supported both military and civilian users needed better observations. Various agencies worked together to develop ground-based solar observatories throughout the 1960s and 1970s. Having observatories with telescopes, often near tracking stations, brought space weather forecasters more of the data advantages that meteorologists enjoyed. With the observatories came another aid to forecasters—better ground-based technology.

Having twelve people look at the typical white-light image of the Sun was marginally better than having six people look at that same image, but increasing the number of people watching the Sun did not help forecasting enough. A white-light image of the Sun from a telescope was wonderful—in 1610. But scientists really needed many different ways of seeing the Sun: for instance, twelve people looking at twelve different views of the Sun.

In 1898 George Ellery Hale built a spectroheliograph, a device that could observe specific wavelengths of the solar emissions. This device revealed what elements exist at the various layers of the Sun. The ball of gas that is the Sun is layered with increasingly hotter regions all the way to its center. Each layer is made up of specific elements that emit light at specific wavelengths, and by observing the different wavelengths of light, scientists can have revealing looks at the Sun's activity (Fig. 5.1). Observers could "tune" the spectroheliograph to receive only one wavelength, say hydrogen-alpha (H-alpha), and then record an image of the Sun. The resulting image showed more than just

Figure 5.1: *Both satellite and ground-based technologies can image the Sun in different wavelengths of light. (Images from NOAA, ESA/NASA, Kitt Peak National Solar Observatory, Big Bear Solar Observatory)*

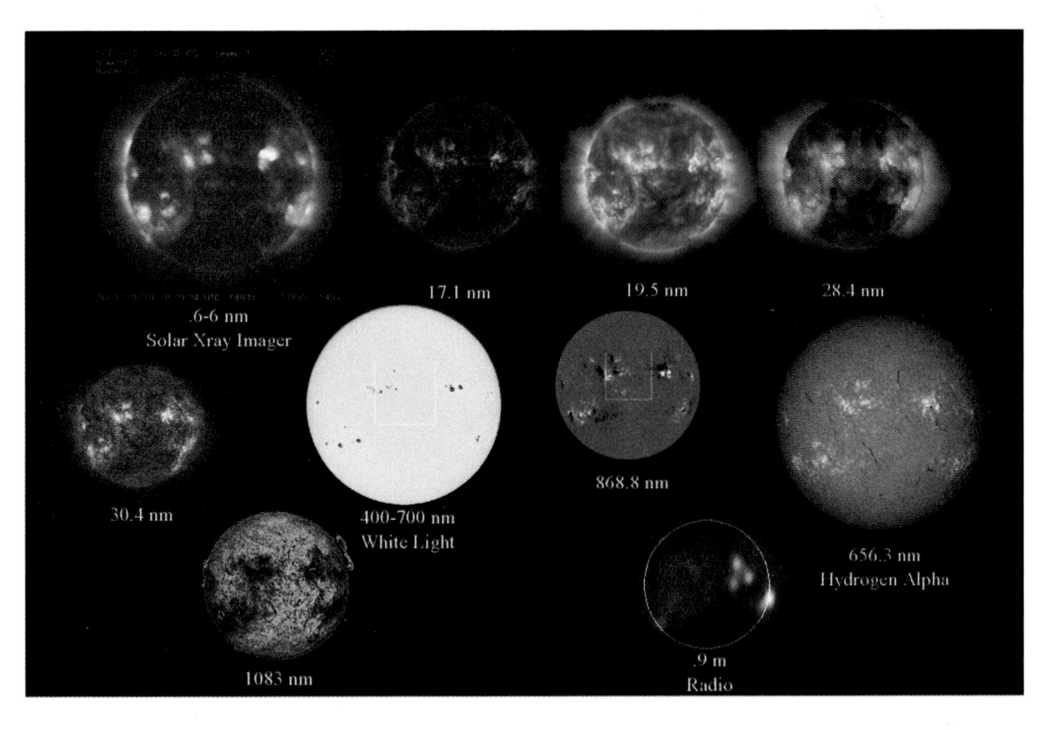

sunspots; there were lines of dark streaks (filaments), very bright spots with rough edges (active regions), and huge explosions of matter from the edge of the Sun that looped back to the surface like the arc of a water fountain (prominences). A similar though simpler technology, putting wavelength filters in a telescope, achieved the same result but was not nearly as precise or clear as the spectroheliograph. However, either of these technologies gave scientists a large number of views of the Sun, a decided improvement over the old white-light image.

The construction and operation of the network of ground-based solar observatories illustrated yet another instance of interagency cooperation. These observatories greatly increased human knowledge of the Sun and allowed for the refining of solar forecasts. Many current veterans of space weather research got their start as observers at these stations—and according to stories they tell, some of them even had a good time. In 1965 NASA established the Solar Particle Alert Network (SPAN) to provide warnings of hazardous solar particle events. It installed this network of telescopes around the world so that the Sun could be observed at all times. NASA and U.S. Air Force (USAF) facilities housed the telescopes: in Houston and Boulder, continental United States; Culgoora and Carnarvon, Australia; Tehran, Iran; Ramey, Puerto Rico; Athens, Greece; Canary Islands, Spain; Palahua, Hawaii; and Manila, Philippines. To aid in this Sun-watching endeavor, the USAF Air Weather Service and national and international observatories joined together to create a Global Solar Flare Patrol network in 1968. The Space Disturbances Forecast Center (SDFC) took on the central processing function of receiving and making sense of the huge amount of worldwide data (Fig. 5.2).

The telescopes were designed and built by the Razdow manufacturing company in Newark, New Jersey. The Razdows allowed the observer to look through an eyepiece at the Sun, capture the image on 35-mm film, and (or) display the image on a TV monitor. NASA contractors initially ran the telescopes, but they knew little about solar physics and merely snapped the pictures. NASA quickly realized the need for trained solar observers, so NOAA and USAF personnel subsequently took over the staffing of the telescopes. A few lucky young observers received their training from the few older experts in the field. Perhaps the most colorful teacher was Helen Dodson Prince, director of the McMath Solar Observatory and a longtime expert in solar observing (Fig. 5.3). She was a professor at the University of Michigan in the 1950s, a time when discrimination against women was common in astronomy. But her students loved her. Prince admonished her trainees to do their jobs well because any one of them might be the only one to see a solar event that everyone else missed. A pioneer in

```
NNNNUUUV
HAUS BOU Ø1174Ø
FROM SPACE DISTURBANCE FORECAST CENTER ESSA BOULDER COLO
SDF NUMBER 248A ISSUED 17ØØZ 1 MAY 1969
CLASS M FLARES ARE EXPECTED IN THE NEXT 6 HOURS.
A. OPTICAL ACTIVITY HAS BEEN MINIMAL SINCE SDF 247B. THE REGION
CURRENTLY AT NØ8W19 CONTINUES TO DEVELOP, AND ITS MAGNETIC
CONFIGURATION MAY NOW INCLUDE A DELTA IN THE SPOT GROWTH REPORTED
YESTERDAY. FLARE ACTIVITY HAS BEEN MINOR. A NEW REGION IS ROTATING
ONTO THE DISK AT S11E81, BUT SURGES AND CORONAL RAIN INDICATE MORE
TO COME. THIS REGION IS VERY BRIGHT ON THE 9.1 CM RADIO MAP, AND IS
THOUGHT TO BE THE SOURCE OF THE MANY DEKAMETRIC RADIO EVENTS OBSERVED
IN THE LAST SEVERAL DAYS, AS FOR INSTANCE BEGINNING Ø1/ØØ57Z
SLIGHT ENHANCEMENT OF PROTON FLUXES AT SATELLITE HEIGHTS HAS BEEN
REPORTED IN THE LAST 12 HOURS. THE GEOMAGNETIC FIELD IS QUIET.
B. CLASS M FLARES ARE EXPECTED FROM THE NØ8W19 REGION. CLASS M
FLARES ARE EXPECTED FROM S11EL COMPLEX, AND THIS REGION MAY POSE
THREAT OF CLASS X EVENTS IN FUTURE. THIS REGION IS CLOSE TO THE
LOCATION OF THE LATE OCTOBER PROTON EVENTS.
C. FLARE AND PROTON EVENT PROBABILITIES FOR THE NEXT THREE 24
HOUR PERIODS BEGINNING (1) MAY/17ØØZ ENDING 4 MAY/17ØØZ.
CLASS M OR GREATER 3Ø/8Ø/8Ø
CLASS X 15/2Ø/2Ø
PROTON EVENTS 15/2Ø/2Ø
D. OTTAWA 1Ø.7 CM FLUX TODAY IS 123. PREDICTED 125/125/135/
E. KP ESTIMATED FOR 12ØØ-15ØØZ WAS 1. KP FORECAST FOR NEXT 24 HOURS
PER KP 15.3.2/2/2/2/2/1/1/1/.
SOLTERWARN
BT
```

Figure 5.2: *An example of a solar disturbance forecast from 1969, distributed by teletype. (NOAA)*

solar physics, Prince had come to Boulder in 1956 to give a talk titled "Birth of an Active Region" and to meet with the solar physics community there. She knew the value of observing and believed that observations made by thousands of people were the key to unlocking the Sun's mysteries. She reminded her students that the Sun had a way of making people humble. The Sun, she noted, would invariably change unpredictably just when they thought they understood it and tried to predict its behavior.

One of Prince's students in Michigan, Joe Hirman, would begin a long space weather career in the SPAN observatories. Hirman had the good fortune to spend a couple of years with Gary Heckman at the Canary Islands Solar Observatory. About the same age and with similar interests, they became lifelong friends who would later make significant contributions in their careers at the Space Environment Laboratory (later SEC) (Fig. 5.4). The Canary Islands Solar Observatory (call letters CYI), like other SPAN stations, was remote and the job fairly routine. The experience of scientists at the Canaries was pretty typical. Gary Heckman often entertained people with stories of his misadventures in the Canaries, and the tale of the goat and the Volvo (perhaps *not* a typical experience for scientists) became a favorite. Hirman recalled the story:

Figure 5.3: *Helen Dodson Prince (center), pictured here at the McMath Solar Observatory, was an extraordinary teacher at the University of Michigan and became the director of the observatory—both unusual achievements at the time for a woman in science. (Emma Ruth Hedeman)*

Life at the observatory involved long drives to and from the observatory at the south end of the island, and our place of residence at the north end of Gran Canaria. It was only about 40 miles [64 km] (the island is 30 miles [48 km] across), but the roads were windy, narrow, dirt, and crowded with animals. Gary, known for running late, was rushing to reach the observatory before sunrise (we needed to have the solar telescopes pointed so that the observations could start immediately once the Sun rose out of the ocean), when he "met" a goat going more slowly in the opposite direction. That goat, the only one the farmer owned, was deeply planted into the frame of Gary's Volvo. Besides the problems at hand—dealing with an irate peasant farmer, a dead goat, and subsequently a late opening of the telescope—Gary had trouble convincing the U.S. Customs agents in Houston when he brought the vehicle to the U.S. that he was not trying to smuggle something into the U.S. The agents actually took the Volvo apart looking for something that never materialized. For years afterward, Gary reported that pieces of the car would suddenly fall off when driving as a result of the incorrect rebuilding of his car.

Gary Heckman returned to the Space Environment Services Center (SESC) in Boulder in 1970 to help NASA with modeling the radiation in space. SESC had taken on the responsibility for interpreting SPAN and other data, running models, and making predictions for the Solar Radiation Analysis Group (SRAG) at mission control in Houston. Joe Hirman left the Canaries in 1972 to run the forecast desk at mission control in Houston in support of Skylab. He and his team monitored radiation levels for the astronauts, trained NASA scientists and astronaut staff in solar observing, and provided twenty-four-hour–a-day support for the Apollo Telescope Mount (ATM) scientists at Skylab.

When Hirman left the Canaries, the new chief observer, Richard Przywitowski, arrived at the station and quickly adapted to the mild, sunny climate of his new post. He had come from NOAA's South Pole Observatory (as had Hirman seven years before) with hundreds of pictures of the aurora australis and surly men with frozen beards. About the same time, the summer of 1972, two University of Colorado students, John Allen and Joe Sullivan, arrived at the Canaries from the Boulder observatory, both well acquainted with the routine of watching the Sun (Fig. 5.5). Besides their solar observing duties, the students made a side trip to Johnson Space Center in Houston for training as part of NASA's Radiation Support Team for the upcoming Apollo 17 and Skylab missions. The Canary Islands Solar Observatory was located near NASA's Canary Islands satellite tracking station, so the students attended several

Figure 5.4: *Gary Heckman (left) and Joe Hirman (right) forged a lifelong friendship during their stay at the Canary Islands. They eventually ended up working together at the SEC in Boulder. (NOAA)*

Figure 5.5: *From left to right: Joe Sullivan, John Allen, and Richard Przywitowski, who manned the SPAN telescope at the CYI observatory, "observe" the Sun while enjoying the warm, safe waters from their vessel. (John Allen, NOAA)*

meetings about the missions just by walking up the hill. From the Canaries, Allen and Sullivan kept a close, excited watch on the Sun for signs of radiation hazard for the Apollo 17 astronauts during the second and third weeks of December 1972. Luckily, the mission passed without any threat from the Sun.

Another observer working in the Canaries was a Spaniard, Félix Herrera. On his first solo observing watch, he witnessed a major flare. Startled and very excited, he called the Boulder labs to report the significant event. Even though he was fluent in English, Herrera slipped into Spanish and spoke very fast when he got excited. It took a long time for the Boulder forecaster to slow him down enough to figure out what he was saying.

When the observatory first opened, it was staffed in two day shifts, seven days a week. It had a Razdow solar telescope with a 15-cm aperture and a Lyot H-alpha filter (Fig. 5.6). Scientists supplemented this optical telescope with a radio telescope. Combining optical and radio telescopes virtually guaranteed good observations of the Sun. If weather obscured the optical observations, the less precise radio observations could be used as a backup. Each observatory had a teletype machine and a Magnafax Telecopier for transmitting black-and-white

Figure 5.6: *The Razdow telescope at the Canary Islands SPAN station (call letters CYI). The station operated for about eight years, from 1966 to 1974. (NOAA)*

photographs by telephone to Houston and Boulder. Color photographs had to be mailed. Shortly after the trained observers arrived, they installed a heliostat (mirrors on a mount that track the Sun) and an optical bench (with focusing lenses and a sunbeam splitter). The telescope, equipped with these additions, moved with the Sun and provided continuous, easily photographable white-light observations of sunspots. Basically the only other necessary equipment amounted to a darkroom, a shortwave radio receiver, and a coffeepot. Starting half an hour before sunrise, the morning-shift observer would open the dome and prepare the telescope. Within the first hour, he would catch up on the solar activity reports received overnight from the other SPAN observatories, make preliminary observations, transmit the morning sunspot picture to the central processing center, and issue the morning solar activity report. As Joe Sullivan remembered it, "Not infrequently, we would receive a 'Good Evening' greeting from the observers in Carnarvon, Australia, who were preparing to shut down for the night just as the Canary Island observers were waking up. Around noon, the teletype might come to life and tell us that the Athens observatory was closing due to smog. The day was devoted to observing the slow-moving Sun and catching bits of world news filtered through BBC External Services on the shortwave."

The Canary Islands observers prepared hand-scaled and hand-typed teletype messages of solar radio flux levels and location and size of sunspots as well as onset time, duration, and size of solar flares. The teletype they used was part of the Astrogeophysical Teletype Network, which was connected to USAF, NOAA, and NASA observatories. The automated system polled each participant every fifteen minutes to collect any coded message that was ready to send. The messages contained brief observations of usual or unusual solar conditions noticed by each station. Preparing a station's message quickly and accurately became a tense task, especially since the message needed to be in international codes (standardized by the International URSIgram and World Days Service) and formatted for international teletype circuits (standardized by the World Meteorological Organization) within the allotted time of fifteen minutes. If an operator had something to report and missed the polling, the message had to wait another fifteen minutes.

Once a day the observers exchanged a summary report of that day's events with other forecast centers around the world. SESC (previously SDFC) collected and brokered the data. Technicians at SESC decoded the data and used the information to generate the first space weather products: alerts and warnings. The products and data consisted of text warnings and forecasts, coded messages, and eventually one solar image a day from each of the observatories, mailed to both Boulder and Houston. Some of the first solar movies taken from space were obtained during the Apollo missions, but they took so long to transmit from the satellite that they were not much use for forecasting. When Skylab went up, scientists finally started receiving transmissions of solar images in near real time.

Joe Sullivan recollected the Canary Islands Solar Observatory during the Skylab era: "During Apollo 17's flight, we were told that no news [of solar events] from a SPAN observatory was good news—solar events could cause mission termination. But when Skylab was up, things were different. The ATM bristled with x-ray and H-alpha telescopes, ultraviolet and extreme ultraviolet spectroheliographs, a spectroheliometer and a white light coronagraph—all highly evolved and precise instruments devoted to solar research. The network of ground-based solar observatories could assist the ATMs and tell them where and when to look! Now SPAN communication became popular."

Shortly after Skylab (unmanned for the launch) went into space on May 14, 1973, a meteoroid shield was ripped off by some unknown object, taking with it one of the two folded solar-power panels. Already severely incapacitated, Skylab met with further misfortune as, once in orbit, the second solar array would not unfold. Without solar panels, Skylab had no power, leaving it dark and cold inside and baking in the Sun outside. Everyone held their breath as this 200,000-pound orbiting laboratory remained without electricity for eleven days, after which the first crew, Charles Conrad, Paul Weitz, and Joseph Kerwin, started its journey to Skylab to assess and repair the damage. The air-to-ground voice communication with the astronauts was broadcast over the public address system of NASA's CYI tracking station so everyone could hear the drama unfold. The connection lasted only for about ten minutes during each ninety-minute orbit. There was laughter on both sides of the Atlantic when the question came down, "Did anyone remember the tin snips?" Careful of the jagged metal, the crew broke the folded array free. Before the crew had left Earth, NASA engineers had designed and built an umbrella out of telescoping fishing poles and nylon that would shade the satellite. The crew deployed this umbrella in space and the satellite soon cooled outside to a temperature that was comfortable for the astronauts and electronics.

While the observers at the SPAN sites focused on their jobs and the space missions, they also kept in contact with the outside world when they could. The scientists at CYI were living under the dictator Generalissimo Francisco Franco—El Caudillo, the Boss—but did their best to ignore that political tension. Late in 1973, however, they received what seemed like catastrophic news that they could hardly ignore. On the morning of Sunday, October 21, 1973, they heard that President Richard Nixon had fired Archibald Cox, the special prosecutor appointed by Nixon to lead the investigation of the Watergate scandal, and that the president had, in fact, entirely abolished the office of the special prosecutor. In response, both the attorney general and deputy attorney general had resigned. Living on a small volcanic island in the Atlantic meant the lifeline to news from the United States was BBC broadcasts from West Africa. Consequently, this news amounted to less than a minute of the five-minute news summary and was followed by reports of scandals in England, Germany, and France. The observers stared at one another, wondering whether their passports were still any good, given the imagined political upheaval in the United States and the dubious stability elsewhere in the world. That same year the outside world made its presence violently known—the observatory was shut down for two days when a sirocco (a West African sandstorm blown out to sea) smothered the island with the Sahara Desert, making the Sun an orange ember in the sky, depositing superfine dust everywhere, and driving everyone into their homes with shutters closed.

Around the beginning of the Skylab IV mission in mid-November 1973, John Allen at CYI realized that the wealth of equipment that they used to watch the Sun every day could also be used to watch heavenly bodies at night. Specifically, he had in mind watching the comet Kohoutek. Allen mounted his Nikon on the solar-tracking heliostat, which they adjusted to track the comet. With help from the other observers, he took four to five-minute time exposures with high-ASA film and produced spectacularly clear pictures of the comet. Comet Kohoutek, which had promised to outshine Halley's Comet (last seen in 1910), failed to achieve such prominence, but the CYI observers and the crew of Skylab IV followed its progress with interest for over a month (Fig. 5.7). The CYI began to send daily pictures of the comet to Houston and Boulder along with the morning solar photographs.

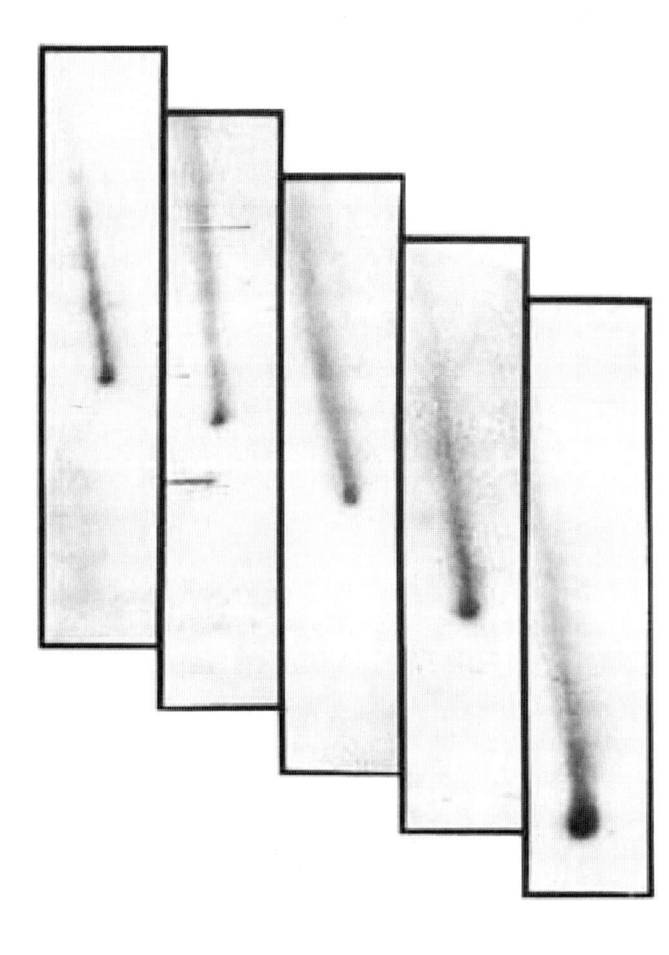

Figure 5.7: *In 1973–1974, Comet Kohoutek was captured by the spectrograph on Skylab on December 13, December 16, January 7, January 8, and January 12. (NASA).*

Although these images were unrelated to solar observation, the amateur astronomers in the group clearly loved their little side project.

NASA scientists arranged to conduct Skylab's missions near Solar Minimum, although the Sun was not very active. The Sun still produced solar events, but since the events typically occurred one at a time, scientists could carefully examine each without the distraction of a very active Sun. The observers at CYI, however, wished they had more to report to Houston. In mid-December 1973, they got their wish. Observatories around the world began tracking a huge filament, a stringy cloud of solar material hovering above the surface of the Sun, as it slowly rotated into profile against the black of interstellar night. On December 19, CYI detected signs of instability in the filament and recommended that the crew of Skylab have the instruments ready. The solar-observing world was ready and watching when this graceful arch of gas (called a prominence when seen in profile) erupted, spewing out light into the black sky. Helen Dodson Prince's directive to be ever vigilant paid off as Skylab caught it all on film (Fig. 5.8). In the picture, the

Figure 5.8: *In December 1973 the crew of Skylab took this famous picture of a huge prominence. (NASA)*

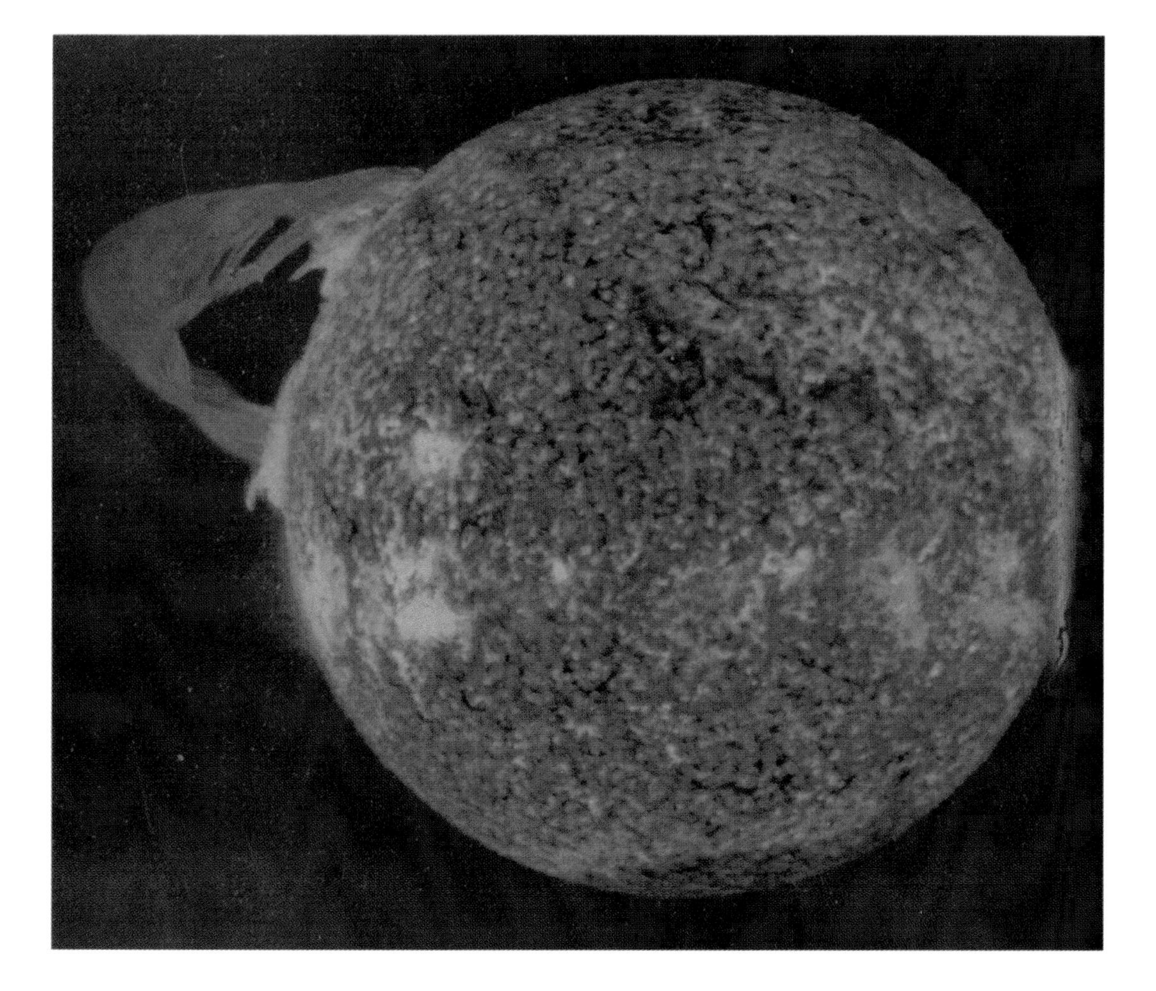

arch was nearly half the radius of the Sun (or the width of twenty-five Earths) and one of the most spectacular solar events ever recorded.

All too soon, the Apollo and Skylab missions ended. They had run their course with remarkable success. In February 1974 the observers began the sad task of dismantling the observatories. Without manned scientific missions in space to support, the fully staffed observatories were unrealistically expensive to maintain, so the telescopes were relocated. They traveled all over the world before they found their permanent homes; one Razdow ended up at the Observatorio del Teide on the island of Tenerife (next to CYI), where Félix Herrera planned to continue his research in solar physics. Other equipment went to various research and academic institutions in the United States and abroad (São Paulo, Brazil; Udipur, India; the Lawrence Livermore National Laboratory in California; and Tehran, Iran). The players lucky enough to be a part of the observing team during Skylab continue to remember and revel in their experiences and the shared excitement of discovery. They received many accolades and recognition for their part in this unusually exciting, enlightening mission.

As easy as it was to dismantle the observatory and its heavy equipment in the Canaries, getting rid of a single teletype machine in Boulder proved nearly impossible. Initially, Space Environment Laboratory (SEL) forecasters had welcomed the technological wonder, especially because getting a teletype machine that was allowed to transmit military messages to a nonmilitary site was unheard of. Although typically secretive, the military wanted the link to SESC, but the functionaries delayed and stumbled through their paperwork to get the machine up and running. SEL workers were thrilled when the "advanced" teletype finally arrived, with strictures on who could receive and read the messages. The AUTODIN, as it was called, was a "black box" that was not to be opened or altered in any way (Fig. 5.9). No one was allowed to move it from one room to another without inspection and certification by Department of Defense (DoD) staff. When the USAF personnel at SEL came periodically to change the codes (password), they would somberly draw the shade over the window in the door before beginning work. Although it was against the rules, SEL staff attached a paper-tape punch to the teletype and eventually a computer, greatly aiding the usefulness of the communications while luckily avoiding the "negative consequences." There appeared to be no enduring paperwork for the transfer of the machine back to military control, yet the annual fee for its use continued to be paid somewhere in the DoD. It took years (at least fifteen, but those not too fond of the machine estimated thirty) to get rid of the teletype after figuring out which military group was in charge of it. The U.S. Army, it turned out, had been paying for it long after

Figure 5.9: *The AUTODIN, a familiar teletype device that received and transmitted solar-terrestrial data and forecasts, was used at SEC for thirty years. (NOAA)*

personal computers and the Internet supplied most of the communication needs at SESC.

The observatory in Boulder, like the one at CYI, had a Razdow telescope with a 35-mm camera that routinely took patrol shots of the Sun throughout the day. During late 1968 the Sun was in Solar Maximum and quite active; the patrol pictures captured striking activity on the Sun and proved useful to the forecasters. The telescope followed the Sun from sunrise above the flat plains east of Boulder to sunset behind the beautiful Flatirons about 3 km (2 miles) to the west. The chief solar observer, Joe Sutorik, examined the sunset pictures day after day and charted the dip of the Sun behind the mountains as it set through the seasons.

One unusual photo, staged for the fun of it and taken by the SPAN telescope in Boulder, would come to have a starring role in the space weather community. Sutorik and one of the student solar observers, Galen McFadyen (who was also a mountaineer), selected a specific date and time to stage a special shot of the Sun. They chose a location along the western ridgeline, accessible to a climber and with an attractive mountain silhouette rising to the right. Sutorik carefully calculated the date that the Sun would set precisely behind that selected location. On December 9, 1968, McFadyen climbed to a point on the top of the ridge and stationed himself looking back down at the observatory. He then waited for the Sun to set behind him. As the Sun slid down along the ridge, McFadyen struck his pose and Sutorik shot the H-alpha picture, catching the figure of the man silhouetted against the huge, detailed Sun. McFadyen said that as the telescope followed the Sun down, he could actually see the Sun reflecting off the red glint of the filter on the telescope below.

Sutorik took several shots that late winter afternoon as the Sun sank behind the ridge. The best photograph became extremely well-known, even famous (Fig. 5.10). The photo has sometimes been printed backward, presumably for artistic reasons or by mistake. Some people over the years could not believe the photo was real, especially because so few

Figure 5.10: *This photograph has been widely published throughout the years by many organizations and publications and has become a sort of "poster child" for space weather in general. (NOAA)*

people have seen an H-alpha image before and because the proportion of the small man to such a big Sun seemed impossible. As digitally altered photos became more common, the supposition, however false, grew that the photo had been a double exposure or otherwise "doctored."

In the 1990s SEC made an effort to re-create the famous picture and produce a negative less worn and degraded than the much-copied original. This time several staff members joined in the climb. Equipped with a radio and coordinated watches, the telescope readied in the observatory, and the date established, the climbers hurried to get to the ridge and find the proper location. Little had changed about the physical landscape since the snapping of the 1968 picture, but it was not familiar ground to the climbers. The first attempt failed to produce any man-Sun picture as the climbers had failed to accurately judge the necessary climbing time. A year later another attempt was made. This time the radio malfunctioned; the star climber did not properly judge when the picture would be snapped and so was not quite ready; the other climbers did not conceal themselves well enough to stay out of the picture, thus adding some extra heads; and finally, the Sun was docile and boring in the sky, nothing compared to the original fiery Sun—and in any case blurry because of very high winds. So much for the advances of technology compared to the art of the original!

Other noteworthy solar observatories exist outside the SPAN sites and main Boulder lab. The National Solar Observatories, McMath being one, were established independently over the last half-century in good viewing areas around the country, dry locations with little air pollution and plenty of sunshine (Fig. 5.11). Universities ran the observatories, and the National Science Foundation funded them. Most of them

Figure 5.11: *The McMath-Pierce Telescope at Kitt Peak National Solar Observatory (NSO) in Arizona is named after the astronomer Robert McMath and the founder of Kitt Peak NSO, Keith Pierce. (Kitt Peak National Solar Observatory)*

still operate today, and many of the best optical images of the Sun have come from them.

The network of ground-based technologies established in the 1960s brought a huge amount of useful data to the field of space weather. The observatories linked together countries around the world and made possible continuous tracking of valuable satellites. Many veteran scientists and forecasters received their training from staffing these observatories. The data gathered at these sites were crucial in making forecasts to warn people of impending solar disturbances. Ultimately, the government agencies funding these sites were not interested in merely collecting data for research purposes, as universities were; they were interested in providing a service to the people. Forecasting space weather disturbances is among the services provided by the Space Environment Center. SEC strives to provide accurate and reliable forecasts, which proves to be a difficult undertaking.

6 Forecasting Space Weather

How often have modern humans dreamed about time travel, inspired either by sheer imagination and longing or by a misunderstanding of Einstein's theory of relativity? Yet space weather forecasters witness a kind of time travel every day. The German astronomer Johannes Kepler observed a supernova as it began its explosion in 1604—but the explosion had happened fourteen years earlier; it had just taken that long for the light, traveling at the speed of light, to reach the Earth. Kepler observed in the present something that was "happening" in the past. Stars that may seem newborn to us on Earth may in fact be billions of years old, as light from them must travel such immense distances to get to us.

The Sun is a mere eight light-minutes away, and so any light that comes to Earth from the Sun left the Sun eight minutes earlier. But since the light we see is in our present, we do not realize the eight-minute delay. As nothing can travel faster than the speed of light, the light from the Sun arrives at Earth before any other solar emission. Particles, invisible to the human eye, leave the Sun at a slower speed and so arrive on a timescale of hours or days. An event seen on the Sun may produce space weather storms of energy and particles that hit Earth from an hour to three days later. The normal solar wind usually travels at only a million miles an hour (400 km [250 miles] per second); light from a flare travels, obviously, at the speed of light (300,000 km [186,000 miles] per second); high-speed particles in the enhanced solar wind travel slower than light at a quarter to a fifth the speed of light (75,000–60,000 km [46,000–37,000 miles] per second). What does this all mean to forecasters? Solar activity hits Earth at varying times, depending on its composition. Although we may know that an event has occurred on the Sun, saying *when* the resulting storm will arrive and how *big* it will be is what forecasters of space weather are expected to do (Fig. 6.1). This turns out to be a difficult thing to do, and forecasters gratefully use every tool at their disposal—techniques of persistence and climatology, supporting staff, satellites, a common vocabulary, and models.

Everyone knows about terrestrial weather forecasting, and the word *forecasting* understandably conjures up a certain picture of well-dressed "experts" standing in front of giant weather maps handing

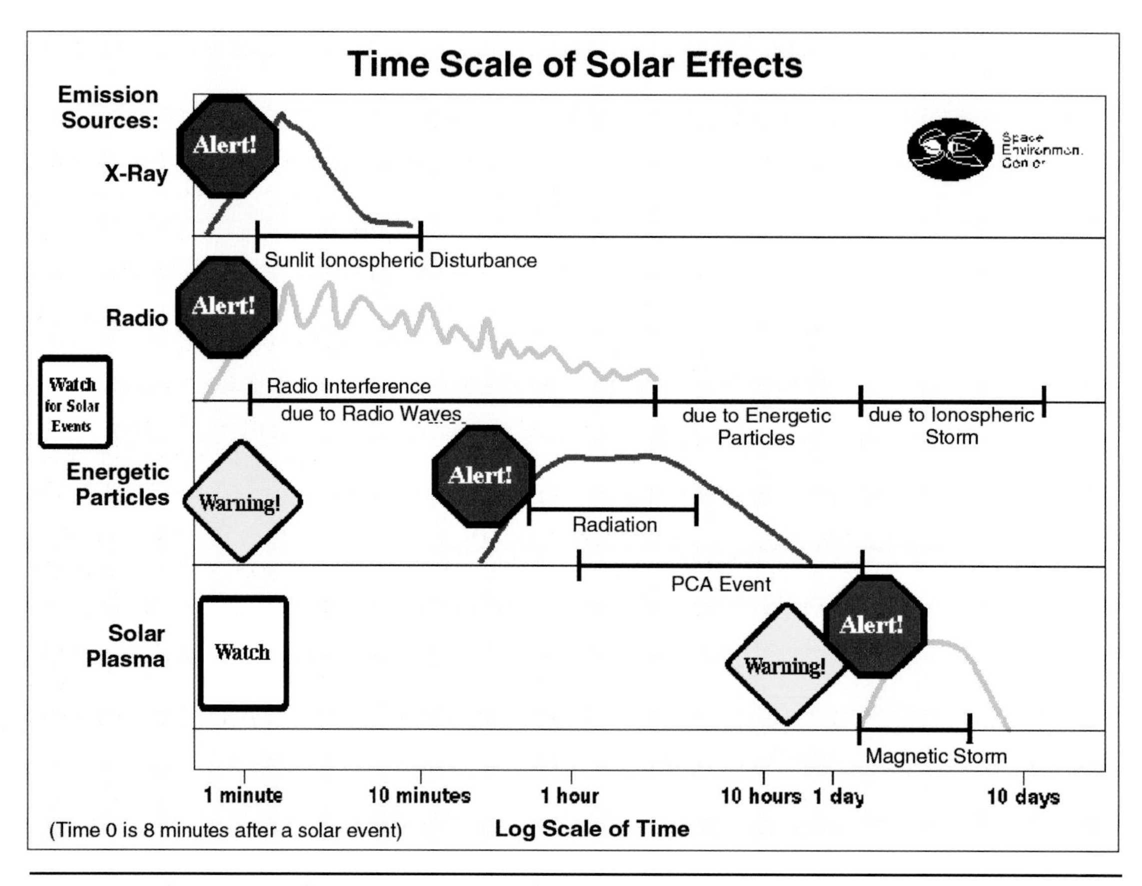

Figure 6.1: *Solar events can have an impact on Earth in minutes or days, depending on the type of emissions and the method of their travel. With slower-traveling emissions, a forecaster has time to issue a warning of a coming storm. (NOAA)*

down nightly weather reports. Space weather forecasting, however, operates on a very different timescale. Events on the Sun tend to evolve slowly and explode quickly. Effects of solar events, as we have seen, hit the Earth over a varied period of time. Joe Kunches, the chief of the SEC forecasting division, attended the annual meeting of the American Meteorological Society in January 2005 and saw just how different terrestrial weather and space weather forecasting are. At the conference, while quite outside the realm of terrestrial weather, Joe gamely gave a space weather briefing. He mentioned and assessed the current day's solar activity, which was rather unremarkable. With nothing of note to mention, he held his audience with short tutorials about solar activity, somewhat wishing for a dramatic solar event. On the third day of the conference, he reported an active region on the east limb of the Sun (this region could not yet be seen directly, so activity showed up as a brightening on the horizon). On the fourth day, Region 720 looked like an active region likely to produce a strong burst from a flare. Joe knew this active region could be of interest in the next two weeks, but, as no one at the conference saw any activity, many meteorologists were not impressed.

Region 720 looked angrier as the days passed, making an event more likely but not inevitable. One week after the meteorological conference ended, on January 20, 2005, a large X3 flare occurred. Having watched the region patiently for a week or more, the forecasters were ready to measure the size of the flare when the blast happened. When the solar light emissions hit the GOES satellite an hour after the flare, forecasters knew a major radiation storm was headed for Earth. When the flare first erupted, forecasters did not know what the effects on Earth would be—would it produce a large geomagnetic storm, massive radio blackouts, or possibly a radiation storm? In this case, the geomagnetic storm was minor, and the radio disturbances were inconsequential, but the flux level (a count of particles) of the solar radiation was the largest of the solar cycle, the largest since the enormous storms of 1989, and the second largest since the satellite era began. "I've never seen the flux levels that high," said researcher Terry Onsager to a CNN reporter about the January 20, 2005, storm. In this storm, the airlines were critically affected, and many had to cancel or reroute their polar flights to avoid the high radiation.

When the radiation alert for the 2005 storm went out, the media responded quickly. But, like meteorologists, reporters are used to the timescale of events such as hurricanes or thunderstorms. For reporting on space weather, the media would prefer a buildup, a "get ready" kind of lead story, but instead they must wait and wait and be on their toes when regions erupt suddenly. Unlike hurricanes and blizzards, space weather storms often start and end in hours and fail to leave behind much lasting damage—this means no lead story that captures the front page for days. Usually only the small space weather community quietly notes damage or losses.

Forecasting space weather begins with the best possible observations of the current conditions on the Sun, including data from the previous three or four days. Forecasters keep a close watch on areas that are growing or shrinking and on the character of the regions that are changing. Looking at the Sun through filtered wavelengths gives a great deal of data, as forecasters want to know what is happening at the levels of various layers. Seeing an agitated, bubbling magma chamber below a volcano's cone could help predict an eruption; it is the same on the Sun. Careful monitoring of the conditions on the Sun tells forecasters whether a solar event (such as a flare or coronal mass ejection [CME]) is likely (Fig. 6.2).

Forecasters can predict the likelihood of a solar event by looking at the amount of condensed energy in the active regions. One can imagine

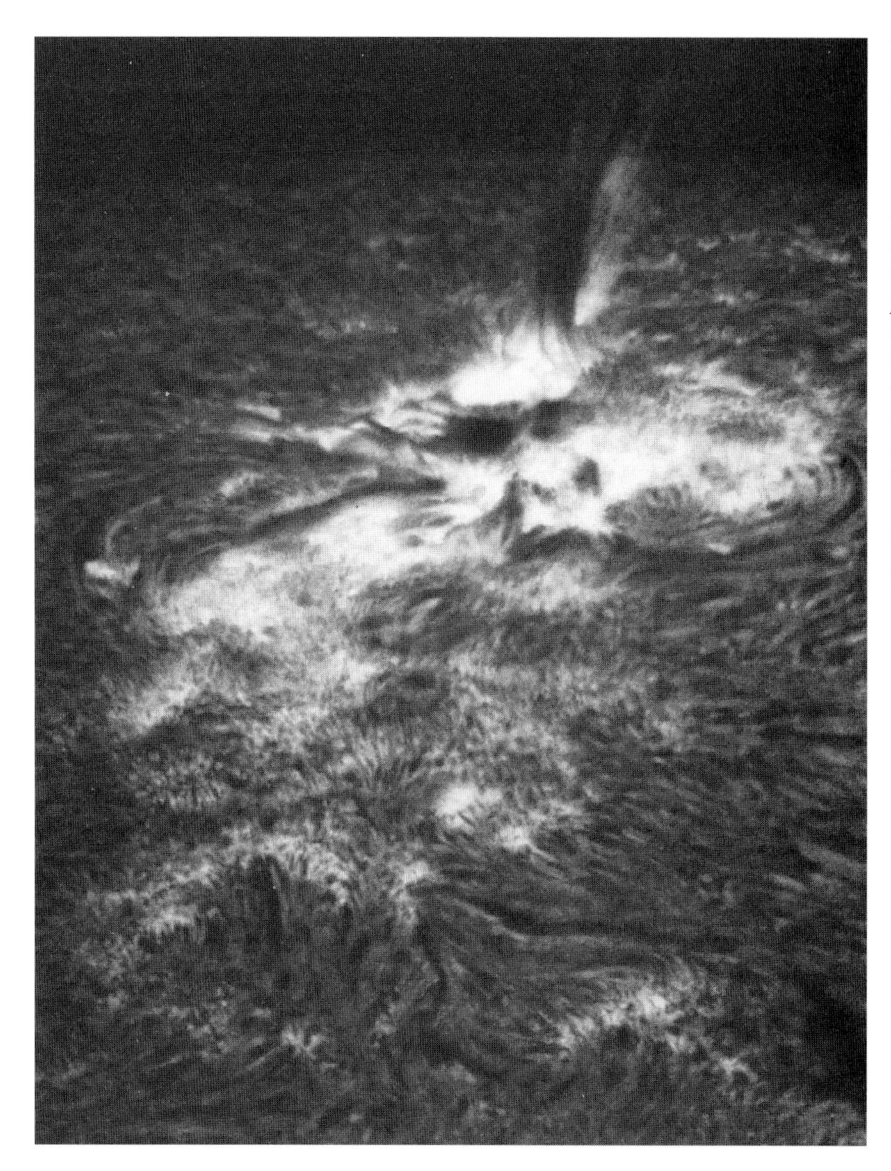

Figure 6.2: *An image of a flare taken from the Big Bear Solar Observatory, May 1970, shown within the highly complex surface of the Sun. The picture evokes a volcano as its plage (French for a sandy beach) hovers over the dark sunspot and the flare jumps off the surface. (Big Bear Solar Observatory, California Institute of Technology [now operated by New Jersey Institute of Technology])*

that meteorologists predict thunderstorms by seeing thick, dark clouds that contain massive amounts of energy in the form of rain, lightning, and thunder. Energy on the Sun is usually "seen" by looking at the characteristics of solar magnetic fields. For instance, if the magnetic field lines are twisted in a tight cluster within a very bright active region, the region may be a candidate for flaring. If the magnetic fields are changing daily and the active region is large, the area is considered unstable and more likely to be the site of a flare.

After forecasters analyze the region for the likelihood of an event, there remain such questions as how large a flare might be and when it would occur. How long will it take before the energy is depleted? If a CME occurs, will it be aimed at Earth? If it reaches Earth, will it cause any storms? When will a geomagnetic storm begin, how long will it last, and how strong will it be? With the data they have, forecasters must try to answer these questions to make a prediction. There are two fundamentals in every forecast: *when* and *what*. To use a simple weather example, you cannot forecast "rain" without saying *when* rain will occur, or say "Tuesday" without telling *what* type of weather will occur on Tuesday. A forecast that gets either the *when* or the *what* wrong is useless. Beyond the simple *when* and *what*, the forecast contains other aspects: how close to the event the prediction is made, how large the event will be, and how long it will last. Another way to judge the quality of a forecast is by looking at the unusual events. If you were, for instance, to predict sunshine in southern Arizona on every day of the year (based on the weather today—persistence), you would, of course, miss the rare snowstorm, but your scores for prediction would be very high. You would be wrong only a few days a year. If you used the average historical conditions for a certain date to predict conditions for the same date this year (climatology, the long-running weather history), you might also be right most of the time. But the value of the forecasts would be low. Both of these techniques, persistence and climatology, help little in more variable conditions. The valuable forecast is the one that alerts us to a hazardous storm.

Thanks to vital satellites and colleagues around the world, the Sun is watched twenty-four hours a day, seven days a week. The Forecast Center at the Space Environment Center (SEC) in Boulder, Colorado, has changed considerably over the years with the increase in technology and international cooperation. Today solar images and banks of data plots showing the Geostationary Operational Environmental Satellite (GOES) sensor readings of protons, electrons, and x-rays fill much of the Forecast Center (Fig. 6.3). Movies of the corona and of coronal holes greatly enhance the full picture of the solar activity that leads to space environment disturbances. Today the SEC processes over 1,400 data streams in real time and distributes data and forecasts in real time that are available anywhere from once a second to once a day (Fig. 6.4).

Interpreting all of these graphs and images requires both forecasters and operations specialists to staff the Forecast Center. Forecasters take the myriad solar data and turn them into a daily forecast. The operational specialists act, in a sense, as the infrastructure for the forecasters. Now listed as physical scientists in the government's classification, they have been called solar technicians, physical science technicians, and communications relay operators over the years. As their earlier names suggest, they are in charge of the communications and technologies required in the forecasting center. While a forecaster may be sitting quietly, deliberating on a space weather analysis, operational specialists

Figure 6.3: *The SEC Forecast Center has changed through the years to keep up with improvements in technology and services. The Forecast Center in the mid-1960s (top two), the mid-1980s (middle), and the mid-2000s (bottom). (NOAA)*

Official Space Weather Advisory issued by NOAA Space Environment Center
Boulder, Colorado, USA

Space Weather Advisory Bulletins are issued when conditions occur that are thought to be of interest to the public.

SPACE WEATHER ADVISORY BULLETIN #05- 3
2005 January 20 at 11:20 a.m. MST (2005 January 20 1820 UTC)

****** STRONG SOLAR FLARE AND RADIATION STORM ******

Active solar Region 720 produced a powerful X7 flare (R3 radio blackout) today at 0701 UTC (00:01 A.M. MST). This is the largest of seven major flares observed in this large and complex sunspot cluster since it emerged as a major flare producer on 15 January.

A strong (S3) radiation storm began soon after this flare. Radiation storms on the NOAA scale are based on proton measurements at >10 MeV. However, this radiation storm is particularly interesting because of the influx of high energy protons (>100 MeV). In fact, this radiation storm, based on the >100 MeV protons, is the strongest since October 1989. A rare, strong ground-level event (GLE) was also observed. GLEs are increases in the ground-level neutrons detected by neutron monitors and are generally associated with very high energy protons (>500 MeV). Elevated neutrons at ground level means there are high fluxes of energetic protons near Earth. High energy radiation storms can be particularly hazardous to spacecraft, and to communication, navigation, and a viation operations at high latitudes.

Active Region 720 is now located near the northwest limb of the Sun; consequently, most of the ejecta from today's coronal mass ejection will not impact the geomagnetic field. However G1 conditions are still possible on 22 and 23 January.

Further major flare activity is possible from Region 720 before it rotates to the far side of the Sun on 22 - 23 January.

Figure 6.4: *The 2000-era Space Weather Advisories are intended for all kinds of users and supplement the more technical alerts. A subscription system lets users choose what products and services they want to receive. (NOAA)*

would be moving from desk to reference shelves, making sure the daily schedules were being met and reminding the forecaster of the various coordination times with the National Aeronautics and Space Administration (NASA) and the U.S. Air Force (USAF). They check and evaluate data, diagnose technical system problems, and do solar map analyses, answering phone calls and getting everything ready for the forecaster to assemble the various pieces for a forecast. They staff the Forecast Center twenty-four hours a day in a three-person rotation. During the quieter times of Solar Minimum, or when funding is low, forecasters are assigned only a single ten-hour shift a day, so operational specialists have all the responsibility for the Forecast Center during the fourteen hours when the duty forecaster is not present. Over the years, with the increase in knowledge and technology, the number of tasks for operational specialists has increased. Undoubtedly, as space weather slowly becomes part of common culture, they will have an even more challenging job.

Over the years the SEC Forecast Center has been blessed with dedicated men and women who have ensured the excellence of space weather services. The staff has included young retirees from the USAF, former observers from the Solar Particle Alert Network solar observatories, National Oceanic and Atmospheric Administration (NOAA) corps officers, and physical scientists trained on local telescopes. So many of the staff at SEC have found the subject fascinating and the work compelling that in 2005 virtually half the staff was over fifty years old. A few years ago Gary Heckman, the senior forecaster at SEC and a recognized veteran of forecasting, took a shift on the forecasting desk. When the regular forecasters miss their shifts owing to illness or unplanned absences, experienced staff members must step in to fill those shifts. What greeted Gary on his shift was a surprisingly huge solar event. He made the regular coordinating call to the USAF to confer and said he wanted to forecast an Ap (a measure of geomagnetic disturbance) of 100. The young sergeant on the USAF desk had never heard of a geomagnetic disturbance that large, and he privately questioned the prediction. But the sergeant had been there less than a year and decided that the SEC forecaster was the expert. In the following USAF briefing, the sergeant announced the unusually high prediction and said that he concurred because "the NOAA people wanted it." Frank Guy, an older USAF solar observer who was well known at SEC, asked who the forecaster was who had made the call. The sergeant said, "Well, that's what worries me. It was some new guy called Heckman." He was, of course, referring to probably the most experienced and longest-tenured space weather forecaster in the world, whose genial personality and status in the community had made him beloved by all. This story was told at Gary's retirement fete. His prediction, by the way, was dead-on.

As a carpenter cannot build a house without tools, a space weather forecaster cannot make accurate predictions without tools such as satellites sending back data about the Sun and the space environment. Each satellite in the forecasters' toolbox contributes different information to help them warn the public of harmful solar activity. A brief reexamination of the satellites discussed in Chapter 4 will give you an idea of what the forecasters have to work with. The GOES satellite has been NOAA's stalwart instrument for years, taking readings in real time of the solar particles, x-rays, and magnetic fields that have traveled 93 million miles from the Sun to the Earth. The data sent to NOAA from this geosynchronous satellite are the most widely used in the space weather community.

Another tool is the Advanced Composition Explorer (ACE) satellite, which provides the only real-time observation of a coming geomagnetic storm. The particles and magnetic field that evolve into a storm travel somewhat slowly, and because they are carried by the invisible solar wind, it is impossible to know without direct observation how big a storm is coming and when. The real-time solar wind (RTSW) instrument on ACE, therefore, is extremely valuable, measuring the orientation of the magnetic field and the speed of the solar wind (Fig. 6.5). If the solar magnetic field happens to be aligned with Earth's magnetic field, little results in the way of a storm. If the magnetic field is aligned opposite to the Earth's magnetic field, there will likely be a huge magnetic storm. Forecasters and users have come to rely heavily on the real-time aspect of the ACE satellite, so that when the satellite is down for even three seconds, people are on the phone to SEC needing to know what the problem is. For people in some fields, seconds can count when relying on accurate space weather data.

The Solar and Heliospheric Observatory (SOHO) is a toolbox in itself, holding a huge variety of solar-watching instruments that can see the surface of the Sun, the corona, the far-flung CMEs, and many other features. Just as meteorologists have come to learn how to read more into their forecasts by looking at the GOES weather photos to determine a chance of rain, space weather forecasters have refined their reading of solar weather by using SOHO images. For example, images from the satellite have allowed forecasters to see a "halo" CME, when a burst of energy is aimed right at Earth that looks like a halo around the blocked-out image of the Sun in the SOHO coronagraph. But a halo might also appear from an explosion at the backside of the Sun, causing matter to travel directly away from Earth. Forecasters can tell the difference only by examining SOHO images of the front side of the Sun for highly active regions and by analyzing GOES x-ray sensor data.

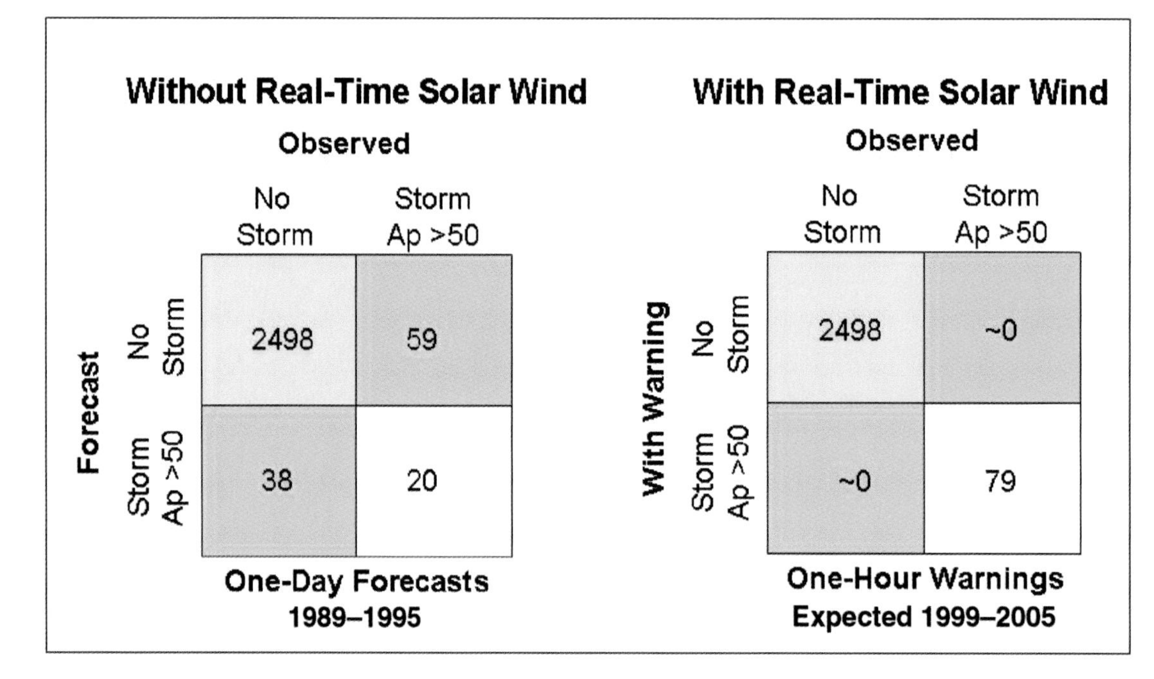

Figure 6.5: *The solar wind "truth test" compares the success of predicting geomagnetic storms with and without a solar wind monitor. The orange boxes indicate failed forecasts. While the solar wind monitor can provide very accurate forecasts, the predictions give only a one-hour lead time, as the data from an L1 satellite take an hour to reach Earth. (NOAA)*

Other data line the forecasters' toolbox (Fig. 6.6), and the space weather community has become ever more dependent on observing satellites. As ground-based observatory networks have shrunk, satellites have become increasingly relied upon. So even though all satellites are vulnerable, scientists plan to launch more that will observe the Sun. Several are on the way; for example, the Solar Terrestrial Relations Observatory (STEREO) mission will look at the Sun from the left and the right of Earth, giving three-dimensional information about emissions from the Sun and the solar wind, and NASA's Living with a Star program will add more solar observational instruments to the collection.

Forecasters, beyond needing advanced equipment, needed a standardized way of speaking about what they saw. Forecasters of 150 years ago needed to identify areas on the Sun, for instance, to recognize an active region as it reappears on the east side of the rotating Sun. In 1863, Sir Richard Carrington established a standard that is still used today for identifying a "zero point" on the Sun to mark one complete rotation (Fig. 6.7). His standard provided a reference that is still used today. The standard allowed observers to accurately watch specific regions on the Sun as those regions evolved. Besides clear locations, observers needed a common language that came from measurable data, not from what the observers saw with the naked eye. Otherwise, one forecaster's "big flare" might be different from what another forecaster called a "big flare." Three developments over the years would help clarify what forecasters observed, both for the forecasting community and for non–space physicists. Each clarifies a different aspect of the solar-terrestrial environment.

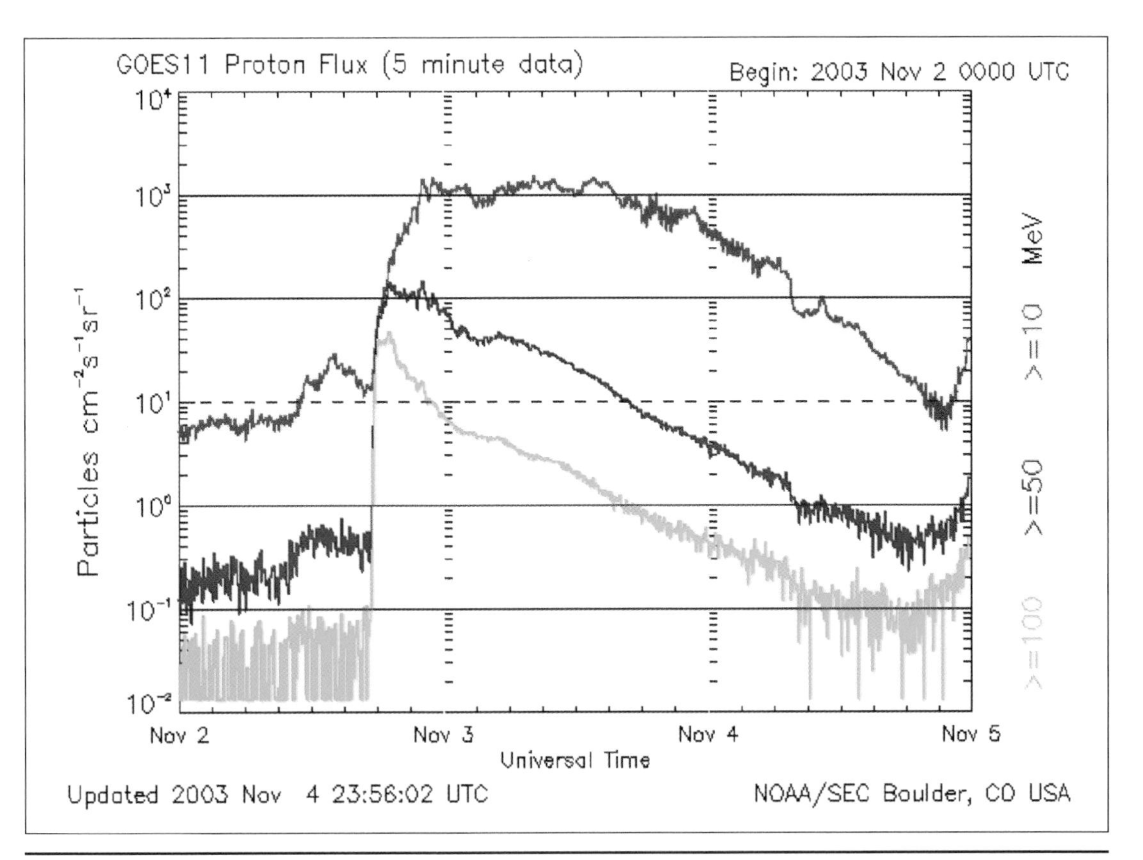

Figure 6.6: *This proton plot is one example of products created by SEC. (NOAA)*

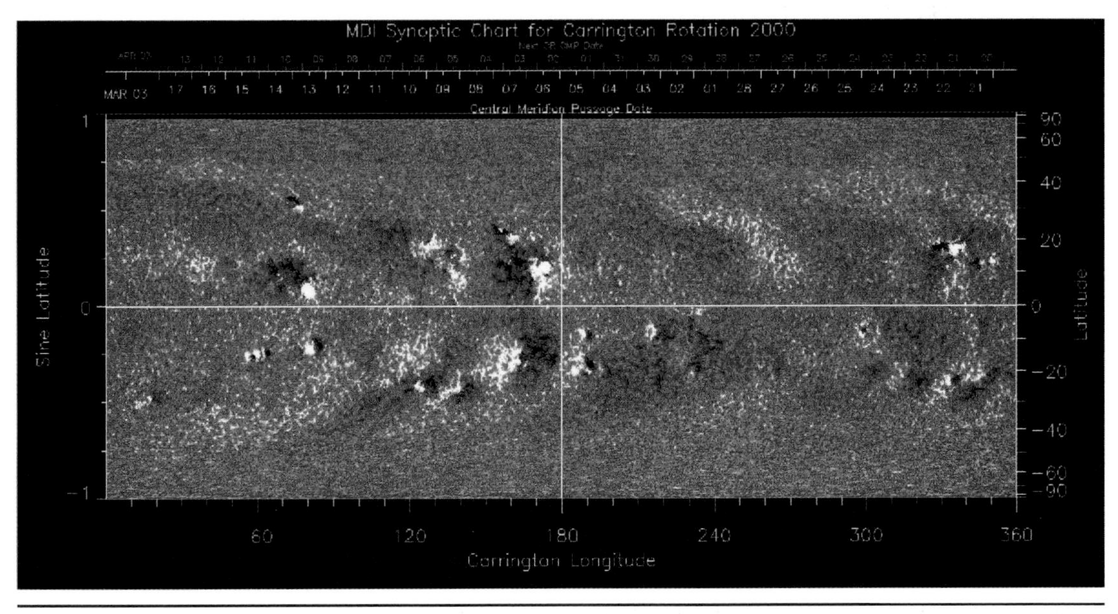

Figure 6.7: *A synoptic map (not unlike an Earth weather map) shows the Sun for one full solar rotation, twenty-seven days. Carrington Longitudes are still used today to coordinate observations. (Stanford-Lockheed Institute for Space Research)*

First, in the mid-1960s Don Baker classified solar flares using x-ray data. The Naval Research Laboratory used German-made Vela V2 rockets, captured by the Americans after the war, to measure x-rays from the Sun. Baker, then working at the Space Disturbances Laboratory (SDL), used these data to identify the wavelengths, 1 to 8 angstroms, that best characterized a flare's intensity. He created NOAA's flare classification, labeled "CMX": C, M, and X stand for small, medium, and large flares, with a range within each category from 1 to 9 (e.g., an M1 flare is one step higher than a C9 flare) (Fig. 6.8). Having C represent the smallest flare left A and B open should there be observations smaller than those currently known. Similarly, Y and Z could follow X if scientists discovered extremely large flares. A and B flares can now in fact be seen with improved instrumentation and are categorized as such. Y and Z have never been used, despite the orderly progression that should have followed as scientists classified bigger and bigger flares (an X1 to X9 would be followed by Y1 to Y9, then Z1 to Z9). Instead we have now seen an X28 flare. This classification system replaced an older one that reported flux numbers; that is, the flare is an M5 rather than flux $Ø \leq 5 \times 1.0^{-05}$ watts/m^2. The improvement was obvious.

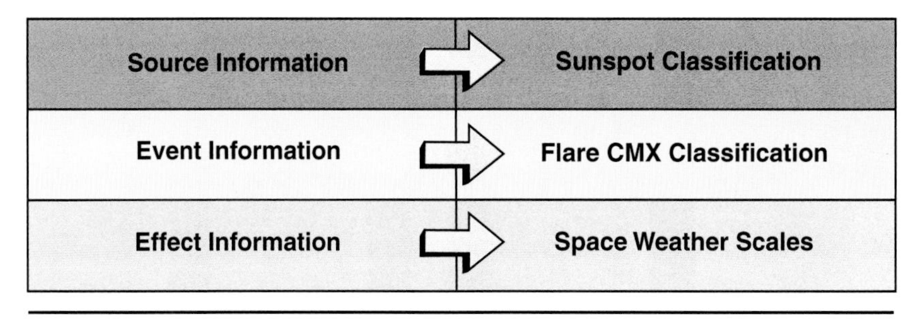

Figure 6.8: *These three classifications of different aspects of the solar-terrestrial environment have improved communication both within the field of space weather and with a wider public audience. (NOAA)*

Second, forecasters needed a shorthand to describe active regions on the Sun. Handmade drawings of sunspots had been done for years, but the drawings had no qualitative information other than what an expert might interpret. By the mid-1960s forecasters needed a better way to discuss and objectively categorize solar features. A young observer, Patrick McIntosh, who had recently graduated from Harvard College, began working at the Space Disturbances Forecast Center. Building upon the modified Zurich Sunspot Classification method (for describing and quantifying the character of sunspots), McIntosh produced a classification system for both sunspots and the active regions around them (Fig. 6.9). Using these lettered codes that describe, among other

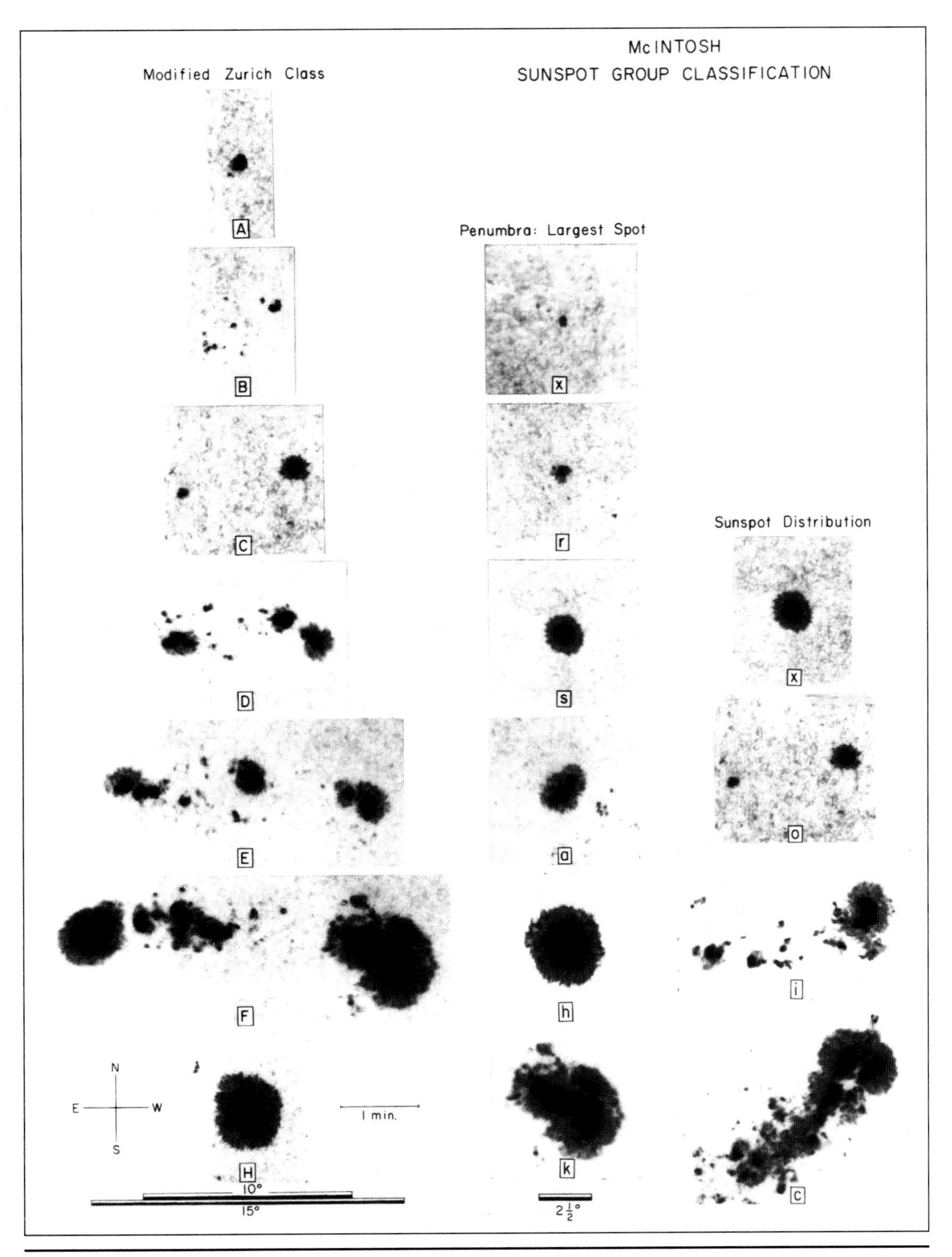

Figure 6.9: *The three-component McIntosh sunspot classification is based on the general form Zpc, where "Z" is the modified Zurich Class, "p" describes the penumbra (grayish area) surrounding the principal spot, and "c" describes the distribution of spots in the interior of the group. There are sixty valid McIntosh classifications (examples: Dao, Eao, Ekc, Fai, Fkc, Fko). (NOAA)*

aspects, the size and shape of sunspots and active regions, forecasters could work out what kind of active regions (what combination of codes) would most likely produce a flare. Although still not entirely objective, this classification gave forecasters, especially inexperienced ones, a systematic and more objective way to describe active regions.

The third classification system was the NOAA Space Weather Scales (Fig. 6.10), designed in 1999, which list three types of storms on Earth as measured by three physical parameters of solar activity. Spurred on by Barbara Poppe, SEC's education and outreach coordinator, SEC staff designed three scales akin to the hurricane scale or the tornado Fujita scale. They classify space weather events so that everyone can understand these events, at least minimally. Each of the three types of storms is rated based on long-used physical measures: geomagnetic storms (G1 to G5) are derived from a magnetic index taken from

Figure 6.10: *The NOAA Space Weather Scales describe the categories of three types of storms that can affect Earth as a result of solar activity. They are a useful communication tool for users and the public alike, although they admittedly oversimplify space weather storms, which are in fact very complex. (NOAA)*

NOAA Space Weather Scales

Category		Effect	Physical Measure	Average Frequency (1 cycle = 11 years)
Scale	Descriptor	Duration of event will influence severity of effects		
Geomagnetic Storms			Kp values* determined every 3 hours	Number of storm events when Kp level was met (number of storm days)
G5	Extreme	Power systems: widespread voltage control problems and protective system problems can occur, some grid systems may experience complete collapse or blackouts. Transformers may experience damage. Spacecraft operations: may experience extensive surface charging, problems with orientation, uplink/downlink and tracking satellites. Other systems: pipeline currents can reach hundreds of amps, HF (high frequency) radio propagation may be impossible in many areas for one to two days, satellite navigation may be degraded for days, low-frequency radio navigation can be out for hours, and aurora has been seen as low as Florida and southern Texas (typically 40° geomagnetic lat.).**	Kp = 9	4 per cycle (4 days per cycle)
G4	Severe	Power systems: possible widespread voltage control problems and some protective systems will mistakenly trip out key assets from the grid. Spacecraft operations: may experience surface charging and tracking problems, corrections may be needed for orientation problems. Other systems: induced pipeline currents affect preventive measures, HF radio propagation sporadic, satellite navigation degraded for hours, low-frequency radio navigation disrupted, and aurora has been seen as low as Alabama and northern California (typically 45° gromagnetic lat.).**	Kp = 8, including a 9-	100 per cycle (60 days per cycle)
G3	Strong	Power systems: voltage corrections may be required, false alarms triggered on some protection devices. Spacecraft operations: surface charging may occur on satellite components, drag may increase on low-Earth-orbit satellites, and corrections may be needed for orientation problems. Other systems: intermittent satellite navigation and low-frequency radio navigation problems may occur, HF radio may be intermittent, and aurora has been seen as low as Illinois and Oregon (typically 50° geomagnetic lat.).**	Kp = 7	200 per cycle (130 days per cycle)
G2	Moderate	Power systems: high-latitude power systems may experience voltage alarms, long-duration storms may cause transformer damage. Spacecraft operations: corrective actions to orientation may be required by ground control; possible changes in drag affect orbit predictions. Other systems: HF radio propagation can fade at higher latitudes, and aurora has been seen as low as New York and Idaho (typically 55° geomagnetic lat.).**	Kp = 6	600 per cycle (360 days per cycle)
G1	Minor	Power systems: weak power grid fluctuations can occur. Spacecraft operations: minor impact on satellite operations possible. Other systems: migratory animals are affected at this and higher levels; aurora is commonly visible at high latitudes (northern Michigan and Maine).**	Kp = 5	1,700 per cycle (900 days per cycle) 2000

* Based on this measure, but other physical measures are also considered. *may change to use other measures, such as DST, as basis.
** For specific locations around the globe, use geomagnetic latitude to determine likely sightings (see www.sec.noaa.gov/Aurora).

NOAA Space Weather Scales, continued

Category		Effect	Physical Measure	Average Frequency (1 cycle = 11 years)
Scale	Descriptor	Duration of event will influence severity of effects		
Solar Radiation Storms			Flux level of ≥ = 10 MeV particles (ions)*	Number of events when flux level was met (number of storm days**)
S5	Extreme	Biological: unavoidable high radiation hazard to astronauts on EVA (extra-vehicular activity); passengers and crew in high-flying aircraft at high latitudes may be exposed to radiation risk.*** Satellite operations: satellites may be rendered useless, memory impacts can cause loss of control, may cause serious noise in image data, star-trackers may be unable to locate sources; permanent damage to solar panels possible. Other systems: complete blackout of HF (high frequency) communications possible through the polar regions, and position errors make navigation operations extremely difficult.	10^5	Fewer than 1 per cycle
S4	Severe	Biological: unavoidable radiation hazard to astronauts on EVA; passengers and crew in high-flying aircraft at high latitudes may be exposed to radiation risk.*** Satellite operations: may experience memory device problems and noise on imaging systems; star-tracker problems may cause orientation problems, and solar panel efficiency can be degraded. Other systems: blackout of HF radio communications through the polar regions and increased navigation errors over several days are likely.	10^4	3 per cycle
S3	Strong	Biological: radiation hazard avoidance recommended for astronauts on EVA; passengers and crew in high-flying aircraft at high latitudes may be exposed to radiation risk.*** Satellite operations: single-event upsets, noise in imaging systems, and slight reduction of efficiency in solar panels are likely. Other systems: degraded HF radio propagation through the polar regions and navigation position errors likely.	10^3	10 per cycle
S2	Moderate	Biological: passengers and crew in high-flying aircraft at high latitudes may be exposed to elevated radiation risk.*** Satellite operations: infrequent single-event upsets possible. Other systems: effects on HF propagation through the polar regions, and navigation at polar cap locations possibly affected.	10^2	25 per cycle
S1	Minor	Biological: none. Satellite operations: none. Other systems: minor impacts on HF radio in the polar regions.	10	50 per cycle

* Flux levels are 5-minute averages. Flux in particles/s·ster·cm². Based on this measure, but other physical measures are also considered.
** These events can last more than one day.
*** High-energy particle measurements (>100 MeV) are a better indicator of radiation risk to passenger and crews. Pregnant women are particularly susceptible.

Category		Effect	Physical Measure	Average Frequency (1 cycle = 11 years)
Scale	Descriptor	Duration of event will influence severity of effects		
Radio Disturbances/Blackouts			GOES x-ray fluxpeak brightness by class (and by flux*) (Wm^{-2})	Number of events when flux level was met (number of storm days)
R5	Extreme	HF radio: complete HF (high frequency**) radio blackout on the entire sunlit side of the Earth lasting for a number of hours. This results in no HF radio contact with mariners and en route aviators in this sector. Navigation: low-frequency navigation signals used by maritime and general aviation systems experience outages on the sunlit side of the Earth for many hours, causing loss in positioning. Increased satellite navigation errors in positioning for several hours on the sunlit side of Earth, which may spread into the night side.	X20 (2×10^{-3})	Fewer than 1 per cycle
R4	Severe	HF radio: HF radio communication blackout on most of the sunlit side of Earth for one to two hours. HF radio contact lost during this time. Navigation: outages of low-frequency navigation signals cause increased error in positioning for one to two hours. Minor disruptions of satellite navigation possible on the sunlit side of Earth.	X10 (10^{-3})	8 per cycle (8 days per cycle)
R3	Strong	HF radio: wide area blackout of HF radio communication, loss of radio contact for about an hour on sunlit side of Earth. Navigation: low-frequency navigation signals degraded for about an hour.	X1 (10^{-4})	175 per cycle (140 days per cycle)
R2	Moderate	HF radio: limited blackout of HF radio communication on sunlit side, loss of radio contact for tens of minutes. Navigation: degradation of low-frequency navigation signals for tens of minutes.	M5 (5×10^{-5})	350 per cycle (300 days per cycle)
R1	Minor	HF radio: weak or minor degradation of HF radio communication on sunlit side, occasional loss of radio contact. Navigation: low-frequency navigation signals degraded for brief intervals.	M1 (10^{-5})	2,000 per cycle (950 days per cycle)

* Flux, measured in the 0.1–0.8 range, in watts/m². Based on this measure, but other physical measures are also considered.
** Other frequencies may also be affected by these conditions.

ground magnetometer readings, solar radiation storms (S1 to S5) are derived from proton flux values measured at GOES, and radio blackouts (R1 to R5) are derived from x-ray fluxes at GOES. Included with the rating for each storm is a list of likely impacts on users. This has been the most helpful in dealing with the public and the media, who always want to know how bad a solar event could be; the scales hopefully discourage such sensational guesses as speculation that the Earth might explode. Consequences are what the media care about, and without the scales they are likely to become confused and exaggerate the consequences. As with the flare classification, the NOAA scales simplify technical science language (120 > 10 MeV protons) into language that more casual users can decipher (an S2 solar radiation storm). During their development, the scales were hotly debated by many members of the staff because the simplicity required to make them effective as a communication tool clashed with the complexity that characterizes solar and space weather events. As imperfect as the scales are, however, they have been a good way to improve communication within the space weather community and with the public.

Well-trained staff, advanced technology, and a universal way of speaking about solar events should lead to accurate solar forecasts. It is not necessarily so. Like solar forecasters, terrestrial weather forecasters constantly struggle with getting forecasts right, yet broadcasters remarkably often change their daily predictions every hour or two without even looking over their shoulders to acknowledge a prior prediction. The public generally writes off the "mistakes" with a shrug and a lingering feeling of distrust in any weather forecast. Those affected by space weather cannot afford to casually distrust solar forecasts. They need to know if the forecasts they receive are any good; if the forecasts are consistently inaccurate, the groups will not use them at all. Space weather service providers are therefore interested in determining whether their predictions are accurate. Verification is basically a retrospective statistical analysis of forecast accuracy.

For space weather, Kent Doggett, a forecaster at SEC, ran an initial verification study in 1993 to evaluate the accuracy of the forecasts. He looked at three types of predictions: (1) likelihood (measured in a percentage) of various levels and types of flares for the next three days, (2) measures of storm size for the next seven days, and (3) next-day probability for an event at a specific latitude. Doggett's study started the ongoing use of measurable parameters of the quality of forecasts over time, thereby allowing one to see the efficacy of forecasts due to any number of complex factors.

A final way forecasters improve their forecasts is by creating and using models. Every field of science creates models to make its observations meaningful, turning streams of numbers into a comprehensible way to see the bigger picture. Physics-based models have allowed the terrestrial weather community to closely predict the precise arrival time and location of hurricanes and many other weather phenomena. The space weather community has been held back from modeling the space environment, partly because of the shortage of observations and partly because the space environment is immense and can be affected by so many external factors. Despite the difficulties, scientists have been developing space weather models slowly over time using all of the data available to them, including past data (sometimes called climatological statistics) (Fig. 6.11). Models might build in such information as whether a region with a certain level of complexity led to a large flare the last time it occurred, or whether it wasn't complex enough to cause much of an event. This would be called an empirical model. Remember, though, that the Sun's behavior is not always "as usual," so past experiences cannot give the full picture.

Computer technology has made scientific modeling possible. For example, a magnetogram shows the magnetic fields on the Sun as dark and light spots, one dark spot (north) paired with one light spot (south). Just these black-and-white pictures can show forecasters the likelihood of solar flares, as the more twisted and complex the magnetic field (seen by looking at the texture and brightness of the spots), the higher the probability of a flare. But a computer-generated physical model, derived using the properties and limits of physics, can extrapolate a three-dimensional image of what the magnetic field lines actually look like (Fig. 6.12). Such a model gives forecasters a more complete sense of how the field lines move, change, and ultimately lead to a flare. The model of the scope of the aurora during various storm levels looks like ovals on a map and is based on a few satellite data points, years' worth of observation data, and a good scientific understanding of how the aurora forms. Physics-based models are the key to precise forecasts, although the limitation of models is that they extrapolate reality rather than replicating it precisely. The SEC website displays several models, as do the websites of a handful of space physics departments at universities.

Most space environment models have been developed in the academic environment. Several space-directed research universities as well as scientists in the Air Force Research Lab and NASA have developed and refined these models. Although the models capture some of the best scientific concepts, they are not always useful in a practical application. The SEC evaluates these models and assesses their value to users. If a model works only when the Sun is quiet, it may be useful as a baseline but ultimately is deemed not useful because it cannot help forecasters predict storms. Some models are highly academic, modeling

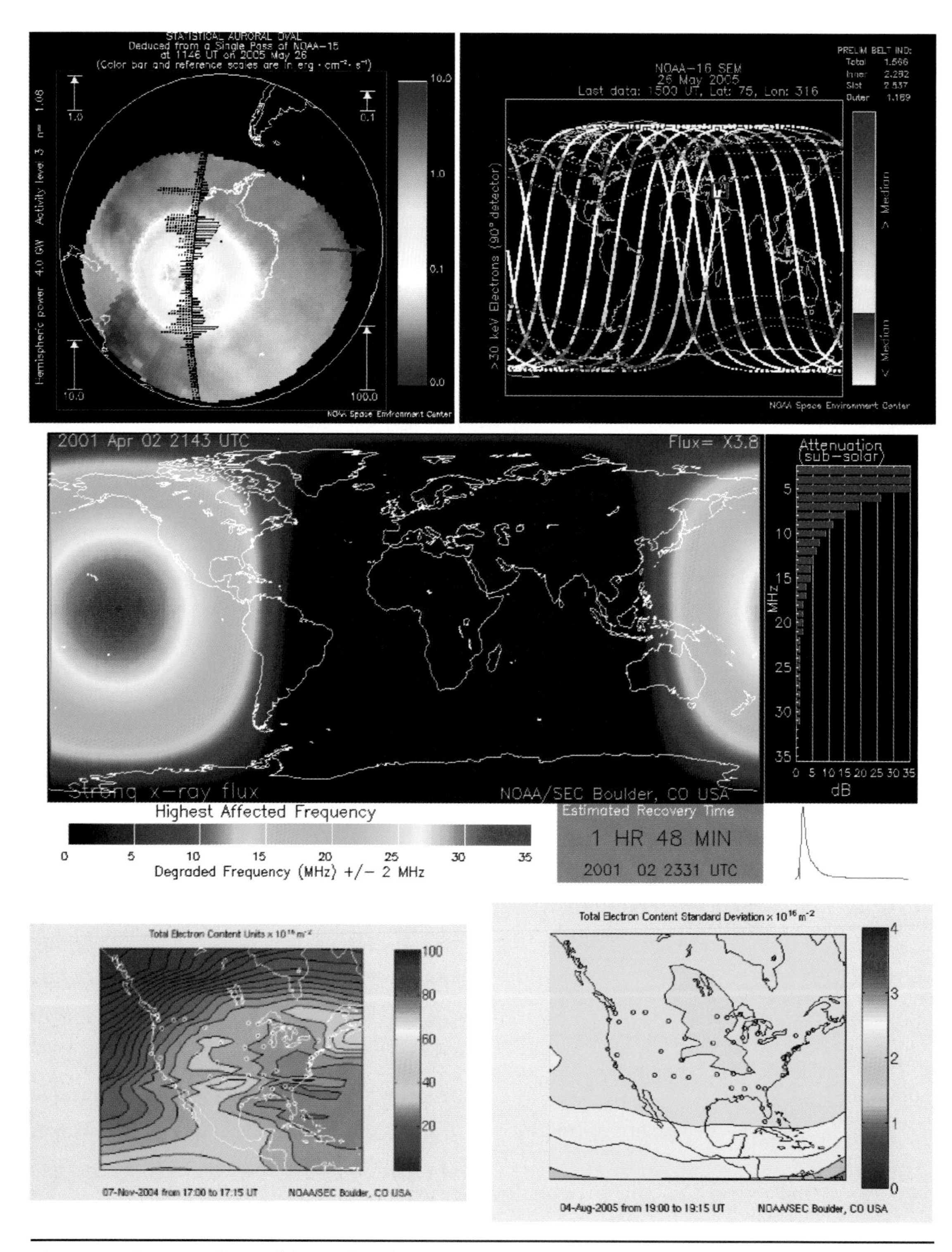

Figure 6.11: *Space weather models are often shown as pictures or movies and let users visualize data. (NOAA)*

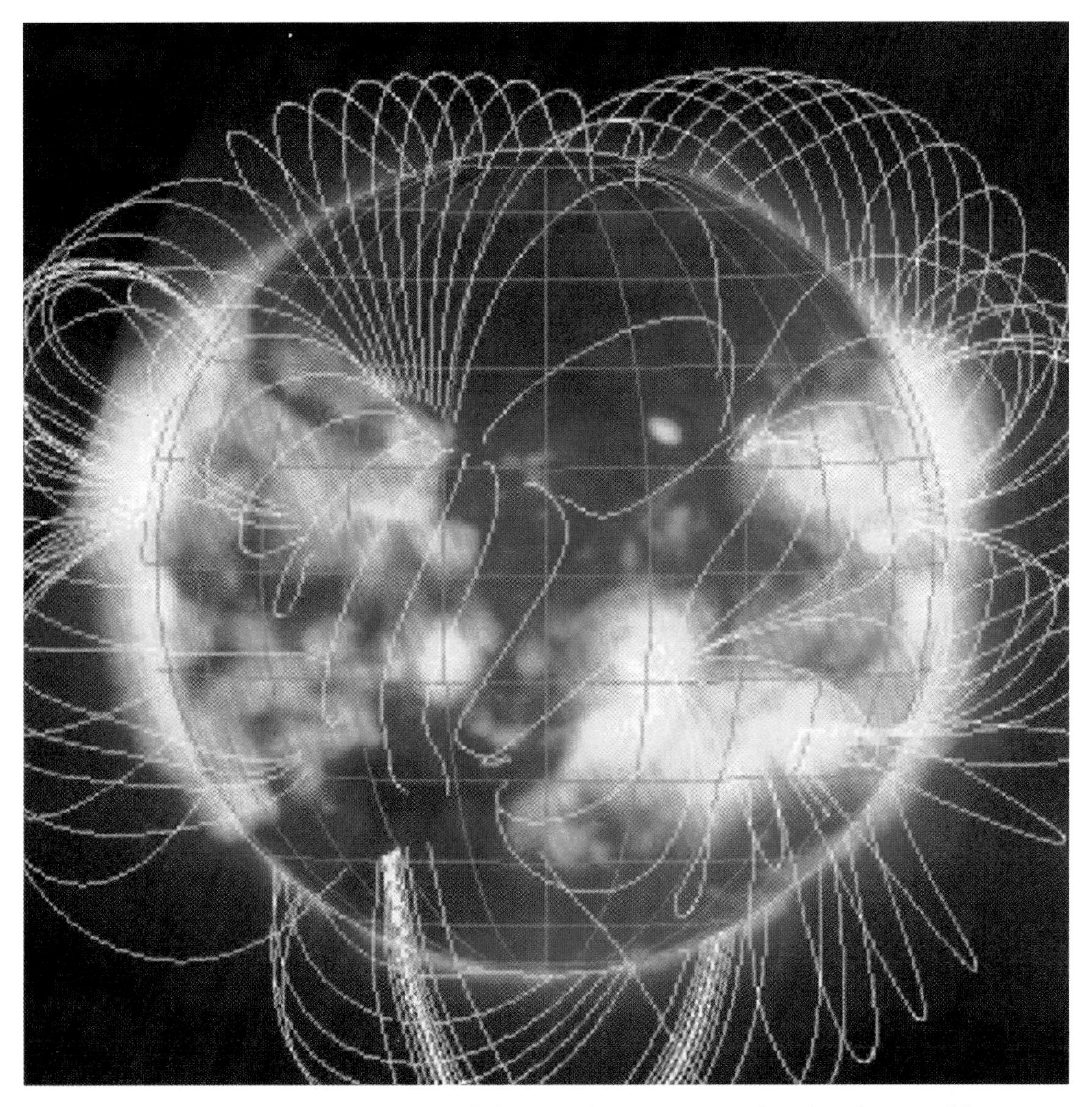

Figure 6.12: *Three-dimensional modeling of the Sun's magnetic field has become a dramatic way of showing the complexity of the solar surface. This model, drawn by Lockheed Martin, can be shown as a real-time three-dimensional user-driven movie. (Lockheed Martin)*

esoteric aspects of the Sun that are not easily related to Earthly events. For example, researchers may be interested in the origin of magnetic fields below the Sun's surface, but if forecasters want a model to help tell a satellite operator when to shut down a system to avoid damage from a storm, it is not worth SEC's time to make the model operational.

Next to the SEC Forecast Center is a room called the Testbed, which contains half a dozen monitors displaying models undergoing the testing process. SEC does not have the time to test every model that is made available to it. It must therefore rate the models for admission to the Testbed: how useful the model is for the user, how useful it is for the forecaster, how easy it will be to take the model through a transition to operations, and how accurate the model is. Once the model is at the top of the list, it moves into operations. The newest and best kinds are assimilative models, which incorporate several types of data into a single model. For example, an assimilative electron-content model of the ionosphere would take into account the density of the ionosphere (related to the number of electrons) as well as a useful measure of GPS signal phase delay. This type of model has not yet reached operation, but several assimilative models are waiting in the wings as candidates for testing.

Despite billions of dollars invested in space-based observations of the Sun, the ability to observe, predict, and warn of impending solar activity is in its childhood. Forecasters have many tools at their disposal—satellites, ground-based technology, models, support personnel, climatological statistics, a common vocabulary—and the quality of forecasts continues to increase over time. But ultimately, forecasters are driven by their mission. Like meteorologists, space weather forecasters provide a service. Weather reporters tell us whether we should bring a raincoat on our camping trip or whether it is a good day to go to the beach. Space weather forecasters help a great variety of people (called users) who are affected by space weather. These users watch the daily space weather forecast, but the stakes are somewhat higher than getting rained on—they could lose millions of dollars. These users provide an impetus to continue to improve space weather forecasting.

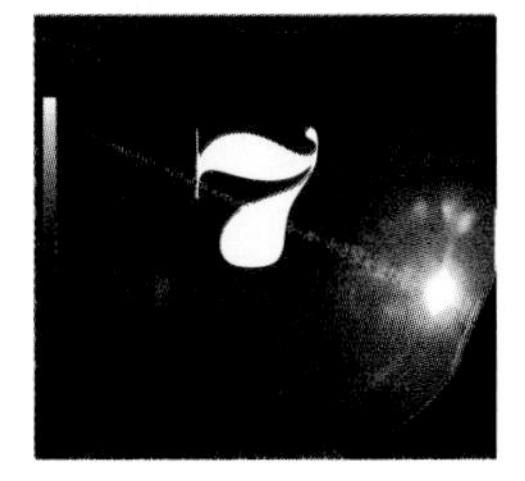

Living with Our Star

Without the user of space weather services, the scientists who study the Sun and the forecasters who interpret its actions would be giving the world something merely academic. After all, who cares whether the Sun is unruly? It turns out that many people care, not in an academic sense but because space weather affects our ability to live and work. In such a technological, global world, space weather can harm the economy and our way of life; as technology increases, so will the risk of space weather (Fig. 7.1). The number of TV broadcasts, wireless communications, and cell phones has begun to grow astronomically in all countries around the world. These technologies— which can be affected by space weather—create the growing number of users.

The term *user* refers to a group or industry that uses space weather data and (or) faces the problems created by space weather. Users are surprisingly varied and range from ham radio operators to the airline industry, from power companies to offshore drillers. The space weather products they receive can be varied as well. Some users want products such as forecasts and warnings, and others just want data. Employees within the government often refer to users as "customers," which implies that they pay for services. As a government agency, however, the Space Environment Center (SEC) provides services free of charge. In recent years, users of all descriptions have benefited from SEC's interest in their needs. The highly interactive annual user conference, traditionally called Space Weather Week, gives users a forum to discuss their service needs. Researchers attend, eager to learn how to gear their research toward functional applications. Forecasters also listen closely to user needs to improve their products. Users learn a lot about the field of space weather.

When speaking about users, it is easy to focus only on the high-visibility users (such as the satellite industry, power companies, the airlines, the military, or the National Aeronautics and Space Administration [NASA]), ignoring or discounting the myriad others. People in smaller industries, such as surveyors and offshore oil drillers, also use space weather products to make their operations more cost-effective. There are numerous users who are hobbyists, such as ham radio operators or

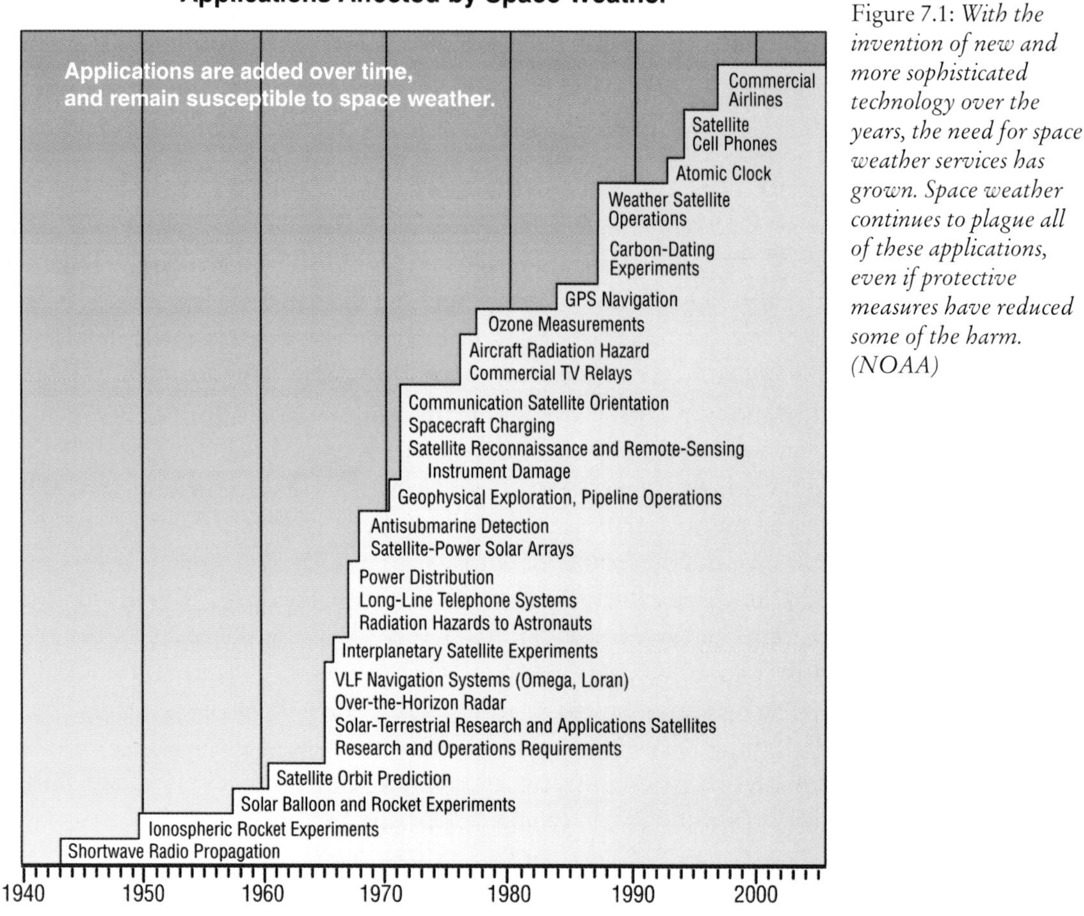

Figure 7.1: *With the invention of new and more sophisticated technology over the years, the need for space weather services has grown. Space weather continues to plague all of these applications, even if protective measures have reduced some of the harm. (NOAA)*

pigeon racers, who notice the impact of space weather and faithfully request SEC services. Forgetting about the smaller users might lead a person to think that space weather has a very narrow effect on Earth, involving just a handful of groups or industries. We will look at the high-visibility groups sorted by industry and the less well-known users sorted by the types of storms that affect them.

Finally, within space weather service there exists another group worth examining: commercial space weather service providers, or vendors. The words *commercial* and *vendor* suggest that the commercial vendors make money by providing services. They would, of course, like to make money, but not all succeed. They are called vendors whether they "vend" or not because they aim to establish a business for space weather services.

Satellite Industry

When satellites were first launched and began operating in low-Earth orbit, the orbital path within Earth's magnetosphere seemed protected from solar interference. More experience, however, proved otherwise. A stormy solar wind can literally push on that protective barrier, lowering it to leave satellites exposed to the highly charged, ionized particles. Several types of radiation can affect satellites, as can storms that cause drag and early "death." But for now, consider just one threat: high-energy electrons. Such particles can penetrate a satellite and leave a static charge on the electronics inside. When the buildup of charge passes a certain threshold, it discharges (like a static shock) and can destroy a computer in the satellite, the satellite's positioning, the solar cells, or the gyros of an orienting telescope. Discovering that this was a problem, the military quickly required its satellites to be built radiation hardened (rad-hardened), that is, shielded from the penetration of ionized particles. However, rad-hardened parts weighed a lot and increased the overall weight of the payload at significant cost, not to mention keeping a heavier satellite in space. The cost of the launch and operation began to compete with the cost of losing the payload.

While the military concerned itself with shielded satellites, the commercial sector entered the scene. In the 1980s, companies began to launch and operate satellites for various users. A few provided satellites for many communication and broadcasting businesses, such as AT&T, CBS, and CNN. Some companies have chosen to purchase and launch a few extra satellites that are light and cheap and to risk losing a few to solar radiation. By the 1990s the demand for rad-hardened satellite parts had decreased because such parts were harder to get and financial considerations had become more important. (Even now military satellites are less robust but cheaper to make and launch. In this case technology has not engineered itself out of space weather danger.) Large commercial industries have reaped the benefits of satellites. In the mid-1990s the commercial satellite industry took off. The emerging opportunities in the global space market drove the demand for satellite assets to exceed $100 billion in 2005 (Fig. 7.2).

All satellites are vulnerable to space weather hazards. Protecting these expensive and valuable instruments can involve an iterative design process. A manufacturer or owner of a satellite may learn, for example, how to protect the solar arrays on the satellite from space weather storms, but instrument designers, who use the satellite as a kind of bus to carry their instruments into space, may call for a higher-voltage solar array so as to provide more power for their instrument; consequently, the satellite designers who draw up the blueprints for the manufacturer must return to the drawing board to create a new, space weather–proof, higher-voltage solar panel, or whatever the change may be. As technology improves, increasingly greater demands are placed on satellites, so designers and manufacturers must constantly alter their equipment.

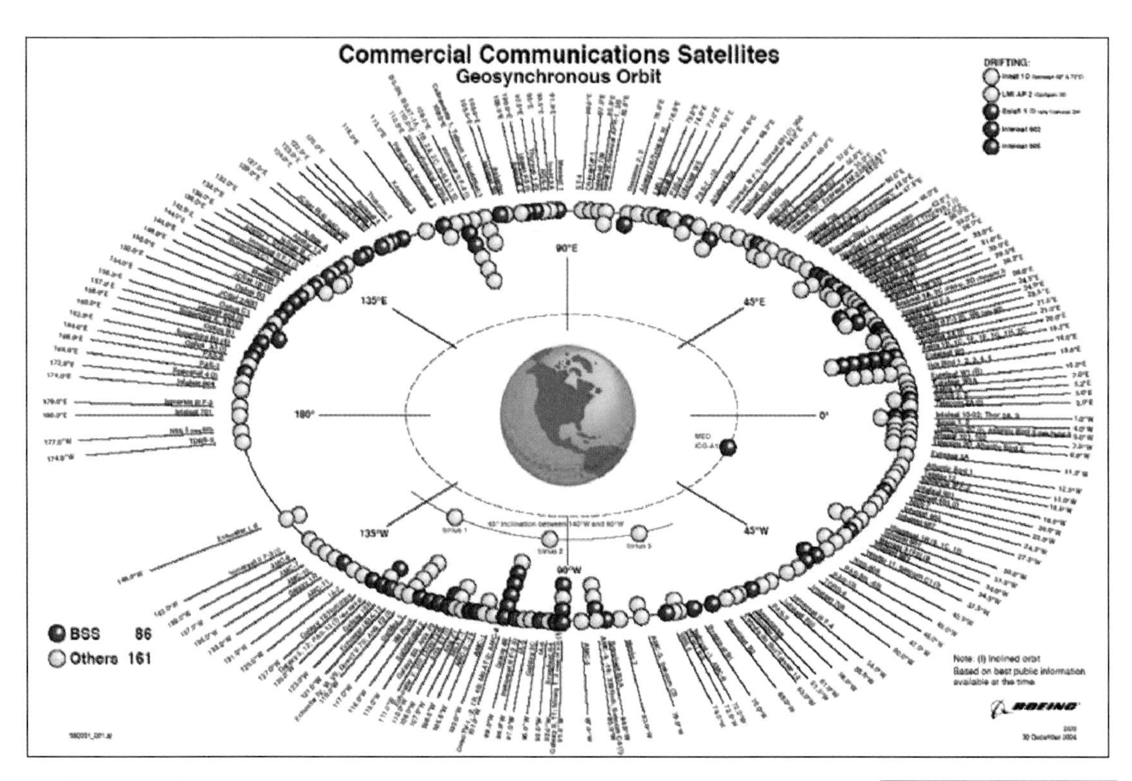

Figure 7.2: *The number of commercial satellites currently stationed at geosynchronous orbit is already reaching its limit. Each geosynchronous satellite has an assigned location in the orbit to avoid collisions or interference with other satellites. The number of satellites shown here is approximate and does not include any military satellites. (Boeing Co.)*

With every change in technology, problems arise that could not have been predicted, so operators must be on their toes to recognize and fix problems.

Space weather is the culprit for many satellite malfunctions. If a satellite's electronics fail, little can be done to fix them, as the cost of sending "repairmen" up can nearly equal or even exceed the value of the satellite. NASA has sent up a few manned spaceflights to rescue a disabled instrument (for instance, one crew installed a corrective lens for the Hubble Telescope), but these flights are extremely expensive and not yet available to private companies. When equipment in a satellite malfunctions, operators on the ground step through diagnostics as to what happened, why, and how it might be fixed. They try to correct the problem from the ground by issuing computer instructions or programming new commands. Design engineers look closely at the failure to gain clues that will help them reduce the vulnerability of their next satellite. Ultimately, satellite owners and operators most strive to prevent such a failure from happening in the first place. If the satellite is about to be hit by solar radiation, as forecasted by SEC, the operator might choose protective action (such as putting the satellite in "safe" mode or shutting it down for the duration of the solar storm). Of course, shutting down the system, whether for a true storm or a false alarm, hurts the customers who pay for satellite service; the total loss of the satellite would hurt the customers more. The goal is to use the forecast to make a decision that creates the smallest disruption to the customers and ensures protection of the satellite.

Space weather can also harm satellite companies economically in a more subtle way. The assets of a company rest on its product investment, and satellites cost a lot. The hint of losing a satellite drives down company stock prices. These financial considerations, with the addition of space weather threatening to destroy the assets in space, make starting up, or investing in, a satellite company a known risk. Understandably, satellite companies are loath to openly discuss satellite anomalies.

Power Companies

Power-generation companies worry about geomagnetic-induced current (GIC). The solar magnetic fields and charged particles traveling on the solar wind couple with the Earth's magnetosphere and set up electric currents in the atmosphere, which can eventually reach the ground, inducing ground currents (usually semidirect or erratic direct current). Direct current (DC) electricity induced on a power company's alternating current (AC) system over which the company has no control creates potential problems. Typically, power companies try to decrease the flow of current through their power lines for the most efficient transmission of power. (They accomplish this by increasing the voltage, which reduces current but is why power lines at high voltage are so dangerous.) When a GIC strikes the power grid, the grid becomes overloaded with current, increasing the power to levels so high as to overheat equipment (Fig. 7.3). A very large geomagnetic storm in 1989 caused the coils in one New Jersey transformer to melt, destroying a $10 million, room-sized transformer bank. When a large storm hits a power-generation system, the entire system can trip out in a matter of minutes.

Hydro-Quebec, a hydroelectric power plant just outside Montreal, supplies the entire province of Quebec with electricity. In March 1989 a huge GIC event shut down the entire plant, and eight million people in Quebec lost power for nine hours. Considering the alternatives, the people of Quebec were lucky. Had the plant been coal-fired, it would have been down for days while the system was shut down for repairs and then brought back up to temperature. Had the plant been nuclear, it would have been down for weeks, for the process of safely shutting down a reactor and restarting it must not be hurried. Had it been summer rather than spring, Quebec might have automatically pulled power from the East Coast grids and collapsed them because of the normally high demand of air conditioners in the United States. The switching of grids to gain enough power happens largely automatically, and in critical times has to be monitored carefully to avoid catastrophic failure. This 1989 storm caused damage and disruption in Canada, and unfortu-

Figure 7.3: *Transformers can literally melt with too much geomagnetic-induced current from a space weather storm. This meltdown occurred in South Africa as a result of the Halloween Storms of October–November 2003. Geomagnetic storms affect the northern and southern hemispheres alike. (ESKOM Generation)*

nately the power grids are now more vulnerable than ever to incidental failure. Since the U.S. government deregulated the power industry, grids across the United States and Canada are now more interdependent on each other and are run with less excess carrying capacity, leaving little room for damaging fluctuations.

Many power companies belong to a power pool, sometimes statewide, sometimes across several states and (or) Canada. These power pools help to regulate energy provision and pass along space weather alerts, as reported by SEC or by commercial space weather service providers. Individual companies can then take protective measures that will mitigate damage or prevent total grid collapse. For example, if a Virginia power company hears of a moderate geomagnetic storm (G2), it may choose to move electric current load away from vulnerable equipment that needs moderate repair. At a G3 (strong) level, it may check with other power utilities to see whether they report any GIC problems, switch on capacitor banks to store up energy, and notify its power station operators. It tries to operate in such a way that current and voltage are as close to in-phase as possible because, as power is the product of current and voltage, the power is strongest when a peak in current is added to a peak in voltage. Companies also try to reduce or avoid performing maintenance on the system at such times. Under certain circumstances, they will reduce the power to some customers. At a G4 (severe) level, they emphatically suggest that their power station operators leave all equipment in service to soak up excess current but not run the system at full power, leaving a "reserve" of power (meaning they have the potential to increase induced current in an emergency). At a G5 (extreme) level, grid operators will also limit interchanges with surrounding utilities in an attempt to prevent the "cascade effect" that has occurred in several other incidental failures.

Power pools make general plans for the companies in the cooperative, but space weather does not affect all stations in a pool equally. Two aspects influence how strong a GIC will be: the location of the power plant with respect to the geomagnetic pole and the preponderance of conducting rock or nonconducting dirt in the ground. Both highly conductive ground and higher geomagnetic latitudes experience stronger GICs. The location of the aurora generally corresponds with the location of high GIC activity, as both are the result of solar particles being funneled to the poles by the Earth's magnetic field and then southward in the magnetosphere based on particle intensity. Nevertheless, power stations across the world remain alert to the potentially global and fast-moving GIC hazard.

Nuclear power plants have also been affected by solar radiation storms: during major solar proton events, monitors that watch the nuclear reactions have sounded alarms owing to the high-energy particle showers that could tip the balance of the very controlled nuclear reaction. Although the threat to the wires and transformers is the same as for hydro-powered or coal-fired generators, nuclear generators need special attention because of their dangerous fuel and the possibility of accidents. Because of the special requirements of reactors, the Nuclear Regulatory Commission pays particular attention to the forecasts. Nuclear power plants require time and care to keep the fission under control at all times. If a plant's managers failed to prepare the reactor for a storm's extra GIC load, they could jeopardize the whole area around the facility. Even minor storms lasting for a longer-than-usual time could cause low-level damage. A meltdown could turn a very expensive power plant into a very expensive toxic waste dump.

Airlines

On January 21, 1976, the announcement, "Passengers of Air France Flight 085 are kindly requested to go to the embarking gate," began for a hundred passengers one of the more famous aeronautical events—the first commercial flight of the Concorde. In its first flight, the Concorde covered 10,000 km (16,093 miles) between Paris and Rio de Janeiro in seven hours and twenty-six minutes. It had taken thirteen years for Air France and British Airways to develop this supersonic aircraft that had two great advantages. The Concorde flew at high altitude (around 60,000 feet)—partly because it could fly faster through thinner atmosphere and partly so it could fly above the slower commercial lanes. In addition, the Concorde could span vast distances without refueling and therefore could take advantage of high-latitude and great-circle routes. (A great circle is the shortest distance between two points on a sphere that lies along a portion of a circle that would perfectly bisect that sphere. For example, the quickest route between New York and Hong Kong is over the North Pole.) However, both high altitudes and high latitudes bring problems for aircraft: more exposure to solar radiation, communications problems, and navigation glitches. The Concorde stopped all passenger flights in November 2003, but more efficient supersonic transports are planned. Supersonic and subsonic planes that provide international travel services will have to take precautions against space weather.

Today commercial subsonic planes can make long flights that cross near the poles, such as from London to Hong Kong, or Chicago to Singapore. Such flights are quicker and cheaper because they travel a shorter great-circle distance and avoid the strong jet streams that slow flights flying at lower latitudes (Fig. 7.4). Currently U.S. air carriers run five or six polar flights a day, and European and East Asian airlines have begun to fly these lucrative routes as well. Although polar flights are profitable on average, a threatening space weather forecast can force an aircraft scheduled to fly over the pole to be rerouted to a more southerly route, sometimes necessitating refueling and a change of crew (who cannot be on duty for more than sixteen hours). Southerly routes will add one to three hours to a fifteen-hour flight and will cost extra fuel and staff, possibly totaling as much as $100,000 per flight. During one space weather storm in March 2001, over two dozen flights were rerouted to avoid the

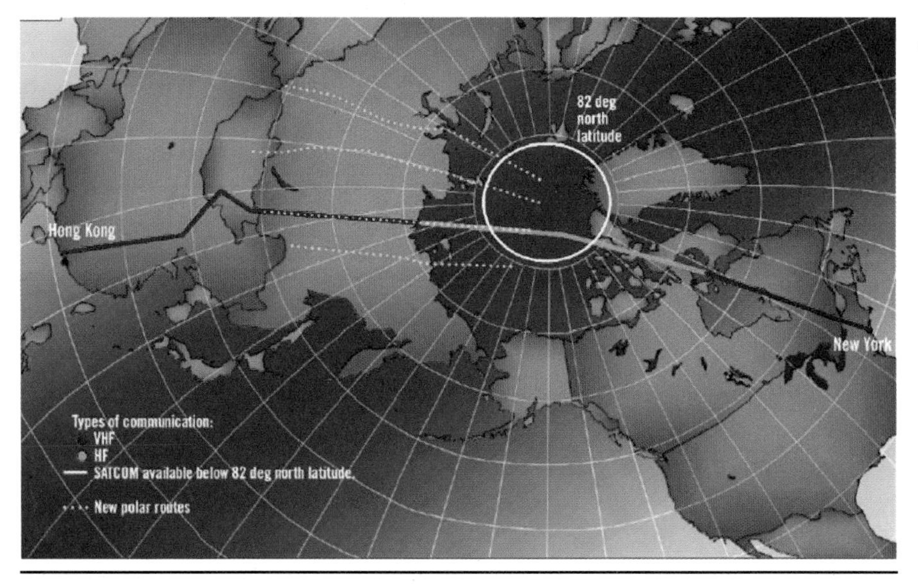

Figure 7.4: *Airlines find many advantages in flying over the poles. However, it is these regions that are most affected by solar activity. Airplanes on polar routes must contend with degraded communications, possible biological impacts from radiation storms, impacts on navigational systems, and a potential impact on electronic systems. (United Airlines)*

poles due to increased radiation levels. The solar storms that occurred in January 2005 caused numerous rerouting and canceled flights, leading one airline company to admit that the five days of storms had cost it a million dollars.

The three big space weather hazards for the airlines are communications disruptions, navigation glitches, and radiation risks. For most flights, normal satellite communications (Satcom) fulfills the Federal Aviation Administration (FAA) requirement that airplanes remain in constant contact with a control center and other airplanes. Satcom uses very high-frequency (VHF) radio links with geosynchronous satellites and so does not rely on radio transmitters on the ground. Based in the equatorial plane in a geosynchronous orbit (22,000 miles above the Earth), Satcom can broadcast only as far north or south as 82 degrees (leaving 8 degrees of latitude out of range). Beyond the northern Satcom limit, airlines must rely on high-frequency (HF) radio communications, requiring ground transmitters and receivers, which in the Arctic are few and far between. Because there are so few transmitters and receivers, conditions must be very good for reliable communication to take place.

Airline communications have always relied on HF radio transmission, but with the advent of Satcom, HF has largely become a backup system for the big commercial jets. As mentioned in Chapter 2, HF communication is both critically important and highly susceptible to space weather disturbances. An HF outage can be caused by two types of space weather storms. One solar radiation storm that occurred on June 14, 2001 (known as the Bastille Day Storm), caused a ninety-six-hour outage in HF communications. The excessive radiation disturbed the ionosphere, scattering or absorbing the radio signals. A geomagnetic storm that occurred on October 31, 2003 (known as the Halloween Storms), caused a 120-hour outage in communications. An outage of more than a few minutes is unacceptable for the airlines, so when extended outages occur, they reroute flights.

The FAA requires tight communication with all airplanes, and that communication is ensured by operators on the ground. They are responsible for relaying all communications traffic to individual airplanes, and they deal with enormous challenges in contacting aircraft, especially over the oceans. The Federal Communications Commission chartered a private company, Aeronautical Radio, Inc. (ARINC), in 1929 to coordinate radio communications for the airline industry. Today ARINC operators monitor air traffic to know where planes are located and use SEC alerts and warnings to prepare the airlines for possible communication failures. For polar-route flights, ARINC monitors for possibly debilitating HF communication conditions and will instruct the pilots to change to higher frequencies to prevent shortwave fadeout. Depending on how bad transmissions are, ARINC will notify New York Air Route Traffic Control, say, that there may be delays in communicating with nearby aircraft. In a worst-case scenario, ARINC will sometimes relay a message to a plane that can then pass the message on to another plane. This practice is not sanctioned but proves to be lifesaving, as it sometimes permits signals to get through to the critical systems.

The second major problem airlines face is satellite navigation. Before the Global Positioning System (GPS) became common in airplanes, airlines relied on an array of guidance systems: compass readings, a check-in system with control towers across the continent, and over-the-horizon radar stations that had limited distances. The Loran C system, run by the U.S. Coast Guard, was established to provide radio-navigation service for U.S. coastal waters. Coverage has been expanded to include the continental United States, most of Alaska, and, thanks to a Canadian/Russian/American partnership, Canadian waters and the Bering Sea. Several problems with Loran C made GPS look better and better. Loran C was accurate to within about fifty to four hundred meters, which left an undesirably large margin of error. Although most pilots had gotten used to the system, it was difficult to use and required complicated calculations. Finally, as with HF communications, space weather storms negatively affected Loran C. As GPS satellite systems began to develop, the accuracy, global extent, and reliability of the equipment attracted many users. Small-plane independent pilots continued widespread use of Loran C through the mid-1990s, but the commercial airlines began to migrate to the more accurate, and more expensive, GPS.

Although GPS proved more accurate and easier to use than Loran, space weather still played tricks with GPS readings. Although satellites send their signal straight down through the ionosphere (not needing to bounce a signal off the ionosphere as in HF communication), the ionosphere nevertheless created two problems for GPS. Charged particles in the ionosphere can cause the signals to be scattered in many directions (in a process called scintillation). With scintillation, the signal fades in and out, causing the receiver to "lose lock" on the signal and ultimately fail to get the data. A disturbed ionosphere can also delay the GPS signal, which will then be inaccurate. GPS calculates ground positions using three measurements: the distance between a known satellite and the ground receiver, the time it takes for the signal to go from satellite to receiver, and the speed of the signal. When space weather inundates the ionosphere with extra electrons, the particles can bunch up and form an uneven medium through which GPS signals travel. The denser patches slow down the signal. Because the signal takes longer to arrive at the ground, the receiver miscalculates the position of the satellite and imagines it farther away than it really is. Unfortunately, the ground receiver and, therefore, often the user fail to realize the error and believe the position reading to be accurate.

To improve GPS reliability, the U.S. Air Force (USAF) developed a highly accurate two-signal GPS system, unavailable to anyone except the military. When it was opened to civilians in 1999, several jumped to use the more accurate and expensive system. The Coast Guard started full operations of its maritime Differential Global Positioning System (DGPS) as soon as dual-frequency GPS became available to the civilian sector. The single-frequency civilian version of GPS had normal errors of 15 to 100 meters. The positional error of a DGPS position, however, was 3 to 5 meters and had the possibility of accuracy to within centimeters. DGPS relied on local beacons that could recalculate GPS readings and transmit the difference between the original GPS signal and a more accurate one. The accuracy of the corrected reading, however, degraded as distance from the beacons increased.

Even more desirable than DGPS was the FAA's Wide Area Augmentation System (WAAS), made operational in 2003. WAAS provided a way to correct any miscalculations in GPS position readings for aircraft. Aircraft trying to land in low-visibility conditions especially needed to know their precise location; a 5-meter error makes a huge difference to a plane trying to land. DGPS worked well for the Coast Guard but could not provide such accurate navigation to airplanes. WAAS can pinpoint locations to 3 meters 95 percent of the time and works anywhere (not just close to a beacon). The FAA was so impressed with the precise accuracy for airlines that it pushed to discontinue Loran in 2000. Congress hesitated to support stopping Loran and funded the continuation of the service "in the short term." Officials feared that relying on GPS alone, without a backup system, would lead to accidents. The vulnerability of GPS was well-known. In 1998 someone at a Department of Defense (DoD) lab near Rome, New York, unintentionally transmitted a spurious signal that jammed the GPS signals in the area. It took several days to trace the source of the problem, and sixteen aircraft reported total loss of GPS over the ten-day period. In that same year a new "toy" went on sale at an air show in Moscow and subsequently on the Internet—a device like a TV remote that would jam GPS signals. As GPS signals are naturally weak, this simple device worked quite effectively. During a time of heightened fear of terrorist attacks, the device conjured up images of a disaster. People also knew of GPS's vulnerability to space weather, which could cause the readings to be off, depending on the space weather conditions, by as much as 20 meters horizontally and 30 meters vertically. Although navigation errors of a few dozen meters make little difference to a plane in flight, precision matters a lot when landing. Dual-frequency (or differential) GPS can help with accuracy, but it often takes longer to obtain readings in disturbed conditions, which decreases the value of the system to airlines.

The airlines understand the need for backing up their GPS signals. The WAAS system is also not reliable enough. In certain cases, high-elevation terrain can mask the WAAS signal. The best alternative, for now, is the Loran system. Loran C currently can be used as a secondary navigation system both near terminals and en route, but unfortunately it does not support the landing-approach phase of flight. The FAA and the Coast Guard are currently working on improving the Loran system to be a backup to GPS, and perhaps, in the future, to correct GPS data. Meanwhile, all three systems—Loran, GPS, and WAAS—remain in operation.

The final space weather problem for airlines is the hazard of radiation. The word *radiation* has a galvanizing effect on people because it is so often equated with nuclear blasts. Solar radiation, however, is less concentrated, more erratic, and not as likely to cause immediate medical problems. In fact, scientists have not been able to precisely calculate the risk. The risk of radiation exposure for people flying in airplanes comes from many factors: the size of the radiation storm, the aircraft type, the latitude and altitude of the flight, the makeup of an individual's body, the length of exposure, where an individual sits in the plane, an individual's history of radiation exposure, and so forth. It is recognized that a young fetus (zero to three months) is most at risk, even at the more protected latitudes (farther from the poles); the risk of getting a disease such as cancer during the formative first three months in a pregnancy is heightened. Although pregnant women worry about the occasional flight, airline pilots and flight attendants experience greater radiation exposure, so pregnant crew members obviously undergo the greatest risk. Flight attendants' and pilots' unions have expressed an interest in determining the radiation hazard. In this age of international business, some passengers are beginning to exceed the exposure of airline employees. Although present-day air travel carries with it unknown risk, the future of air travel—commercial spaceflights (discussed in Chapter 9)—guarantees that radiation storms will heighten the impact on passengers.

Military

The military cares a great deal about space weather because of its fifty-odd systems—including satellites, command communications, weapons, and intelligence—that are susceptible to the effects of space weather. Space-based systems alone make communications, intelligence collection and dissemination, electronic attack (such as smart bombs), and navigation possible. The U.S. DoD spends $500 million a year mitigating space weather hazards to its satellites.

Modern-day warfare has come to rely heavily on satellites, cell phones, and high-tech weapons. Smart bombs find their precise targets with guidance systems that use GPS. Without accurate GPS, "smart" bombs become unguided missiles likely to miss their targets. Political as well as humanitarian success depends on accurate strikes. During Operation Desert Storm in the early 1990s, the unfamiliar desert terrain called for accurate location information from GPS. Without constant positioning assistance, troops in an undifferentiated landscape, perhaps in a sandstorm, can quickly become lost or disoriented; the situation can be paralyzing. Troops need hourly updates on space weather to know whether malfunctioning equipment has been affected by space weather or something electrical or mechanical. The U.S. Air Force, the leading DoD space agency, has a heightened military need to monitor the Sun and its impact on Earth's magnetosphere to ensure consistent use of communications hardware. In one example, USAF weather personnel working in Iraq were told by the commander in chief that a heightened attack on the insurgents would take place over the next five days and that during that time he needed 100 percent radio contact with the fighters. He got it, thanks to careful monitoring and responsive changes in radio frequencies to ensure good contact. During the invasion of Afghanistan in 2002, military technologies almost certainly faced space weather disturbances, as the Sun was very active.

National Aeronautics and Space Administration

NASA has faced a wide range of problems with space weather. The radiation hazard to astronauts is more of a concern than for airline passengers. Although most spacecraft travel is within the protective magnetosphere of the Earth, not all spacecraft are shielded all the time from harmful radiation. One crew aboard the International Space Station was warned of a big radiation storm and told to take cover. They hurried to the Soviet part of the craft, one of the oldest parts and a real "klunker," but built in a pragmatic, traditional way so that it was better shielded (Fig. 7.5). Astronauts face the greatest radiation risk when they venture outside the spacecraft, on space walks or Moon walks (Fig. 7.6). Currently, spacesuits worn by the astronauts outside their capsule do not fully prevent radiation exposure. In the near future, it is doubtful that suits can be engineered to sustain an unsheltered radiation blast while remaining maneuverable and allowing the astronaut to see.

Next to astronaut safety, NASA worries about its satellites. Plenty of space weather events have affected the satellites, but most have been saved, at least partially. NASA must worry about its equipment from launch through the satellite's mission. The agency will not launch on days with a high risk of geomagnetic storms, as solar electrons can interfere

Figure 7.5: *Upon being warned of a particularly strong solar flare, the crew of the International Space Station took cover in the older, more shielded part of the spacecraft (uppermost in picture). (NASA)*

Figure 7.6: *Space weather must be on its best behavior before NASA will allow an astronaut to venture out to fix a crippled satellite. Mission requirements call for the latest update on the space environment before giving the go-ahead. (NASA)*

with the electronics in the launch vehicles. Close readings of the geomagnetic indexes are all part of the launch preparations. Luckily, geomagnetic storms cause little disruption at the low latitude of Cape Canaveral, Florida. During this solar cycle, at a launch site in Kodiak, Alaska, there has been one launch canceled entirely because of space weather.

During the mission, spacecraft have to be watched carefully. Space is teeming with satellites, long-dead probes, and broken craft parts, all of which can cause fatal problems. A significant collision with a fist-sized chunk of debris can destroy a working satellite. Even small obstacles can degrade the quality of a telescope's image, disrupt the sensors keeping a satellite pointed at a star, or damage the overall guidance system. M. M. Shara and M. D. Johnson published a study in 1986 in which they calculated the risk of damage to the Hubble Space Telescope due to space debris. They estimated that a 5-mm fragment would likely hit the telescope once during the expected seventeen-year mission; a 10-cm-diameter fragment had a 1 percent chance of hitting the telescope but, if it did, would destroy it. As mentioned in Chapter 3, Air Force Space Command keeps track of all large space debris in an attempt to protect satellites from damage (Fig. 7.7). Space weather complicates its job by altering the density of the space through which the debris moves, throwing the junk out of predictable and stable orbits. When the debris changes orbit, the likelihood of a fatal collision increases dramatically.

Figure 7.7: *Tracking debris is key in keeping the Hubble Space Telescope and other spacecraft functioning, as a 10-cm fragment could destroy the orbiting telescope. (NASA)*

Ham Radio Operators

The amateur radio community, members of which are often called hams (a colorful nickname with unknown origins), is an enormous group of hobbyists. They constitute the largest group of nonprofessionals who know the most about the Sun's activity. There are hams in almost every walk of life, including congressmen-hams and ham-astronauts who take radios to the International Space Station and "call home" to talk to other excited hams on Earth. One national nonprofit membership organization, the American Radio Relay League (ARRL), claims to have more than 170,000 members nationwide. All told, in the United States there are about 675,000 hams; there are 2.5 million hams worldwide.

Hams watch solar activity eagerly; the ARRL website even periodically posts a solar update bulletin. Many hams find long-distance communication most interesting, so they must bounce their shortwave signals off the ionosphere. But as with all radio communication, they must deal with space weather interference. Some hams use satellites to relay their signals, but space weather can cause disturbances there as well. For hams, however, not all disturbances are bad. Under normal conditions, a ham in Chicago may be able to talk only to someone in New York. Under disturbed conditions, the Chicago ham's signal may bounce all the way to Berlin, allowing for some very interesting conversations. In a way, then, hams consider Solar Minimum to be something like a summer lull for skiers. They eagerly await the return of high sunspot numbers and lots of ionospheric disturbances that will yield amazing, unpredictable opportunities in global transmissions. This enormous user base has come to the aid of SEC on more than one occasion by contacting members of Congress and government agencies that threaten to cut funding for their valuable space weather forecasts and alerts.

Power companies, the airlines, the military, NASA, ham radio operators, and businesses invested in satellites are the major users of space weather services. The following is a short description of some of the less visible users, grouped according to the storms that bedevil them. Note that they are no less concerned and no less financially involved in space weather.

Geomagnetic Storms

Geomagnetic storms quickly spread over the auroral zones, affecting many types of technology (Fig. 7.8). Second to the power industry's concern about these storms is the use of GPS as it affects airlines. However, the airlines by no means hold a monopoly on GPS technology. In this increasingly technological world, GPS has become popular with a range of people (Fig. 7.9). In the last two decades the number of civilian appli-

Four Minutes of a Superstorm - March 13, 1989

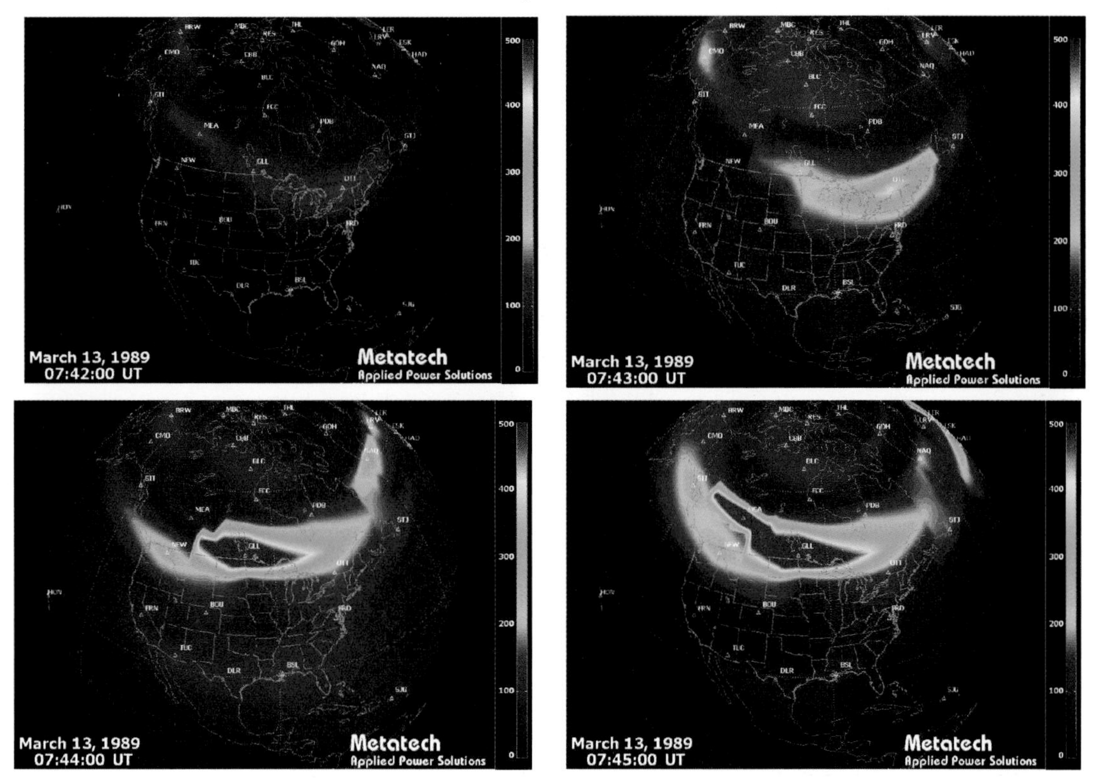

Figure 7.8: *This sequence of pictures shows the evolution of a huge geomagnetic storm. During the March 13, 1989, superstorm, the rapid development and movement of the geomagnetic field disturbance leading up to the collapse of the Hydro-Quebec system happened in four minutes. From calm conditions to collapse took only ninety seconds. (Metatech)*

cations of GPS has soared, especially with the release of dual-frequency GPS by the DoD.

The civilian world benefits from GPS in a great many ways—sometimes in lifesaving ways. In emergencies people call the police from home or a pay phone but often do not know where they are, or hang up before they remember to say. Luckily, within minutes the police can identify the phone number and the physical location of that phone, sending the emergency service vehicle to the exact location. But in the increasingly technological world, a huge number of emergency calls are made from cellular phones. Tracing cell phone emergency calls brings up a host of problems. The greatest advantage of cell phones for ordinary use is that they are not tied to any location. A cell phone with a Colorado area code may make an emergency call from California. Police and other emergency organizations recognize the limitations that cell phones hold and are pressing the legislatures for a solution. The solution is that cell phones must know their location, which they

Figure 7.9: *GPS sales reached $4 billion in 1998 and are expected to skyrocket to $22 billion by 2008. "In war-fighting arenas, GPS has a proven track record. For instance, in the first six days of the U.S.-led Operation Iraqi Freedom, more than 80 percent of the munitions that hit several thousand targets were precision-guided via GPS," space.com writer Leonard David stated. (www.space.com)*

can get from a GPS satellite (if the cell phone has a built-in GPS receiver). In an emergency, then, the cell phone can "tell" the emergency personnel its location (privacy issues factor into the discussion). The cell phone and emergency service industries have been developing an enhanced 911 (E911) system as a response to the expressed need. Connecting GPS to cell phones could save many lives, but the system would not be 100 percent reliable, as both cell phones and GPS can be affected by space weather.

Civilian convenience is also increased by GPS. Toll roads throughout the United States have begun to collect fees using electronic devices attached to a car or truck. Europe is developing a similar system to track trucks traveling on toll roads using the European equivalent of GPS, called Galaxy. A central processing location continuously receives the truck's position and automatically calculates a monthly fee, thus avoiding congestion caused by traffic slowing to pay tolls. The truck's precise GPS locations are needed to identify its route and to clearly tell a toll road (charged by use) from a frontage road (free). An error in pinpointing location can mean major costs, either for the roadway operators or the truckers.

Hydrological and gravity surveys of the Earth require accurate GPS readings. Technicians use airplanes to run tracks along surveying lines to measure gravity or geodetic anomalies that suggest the presence of oil reserves. The surveyors, representing university, government, and corporate interests, check the space weather carefully. If they run a survey during a magnetic storm, the data are considered worthless, as the possible oil site will be incorrectly located. The cost of flying or renting the airplane might be $20,000 and the delay costs due to a space weather storm (including the lost time of the crew and scientists) could mean a loss of as much as $100,000 a day. If a survey must be called off, technicians save the cost of the run, but the work must be delayed until the conditions improve—another kind of loss.

Disturbed GPS can also affect ground surveying. Transportation departments throughout the country, interested in laying precise roads or marking critical junctions, keep an eye on geomagnetic activity that could cause incorrect survey readings. If the activity levels are high, they will call off the survey work for the duration of the storm. NOAA's National Geodetic Survey (NGS) also carries out ground surveying, but for a broader purpose. Once called the Coast and Geodetic Survey, this government organization takes precise measurements of U.S. land and shores. It desires to know the GPS-corrected location to within a centimeter, which requires repeated measurements using dual-precision GPS. During times of poor signal transmission due to space weather, NGS surveyors must take even more measurements and wait a longer time for the calculations that determine the correct location.

GPS's ability to give accurate vertical readings, that is, measuring heights, has opened up several new applications that NGS supports. Farmers (users of NGS data and services) can more efficiently apply fertilizer and water to their fields with accurate elevation maps. Farmers can plow fields faster when GPS height readings can guide blade depth so the farmer can keep the furrows a consistent depth, even on slopes. NGS data have also been used to plan hurricane evacuation routes, monitor erosion and floodplains throughout the country, and check height and slope of terrain in coastal areas (some areas have shrunk a foot in the last ten years)—all thanks to accurate GPS readings.

Offshore oil drilling rigs and oceanographic study drilling ships rely on GPS to give them precise locations. Often these outfits have to remove their drills from the holes they are mining or exploring. For example, if seas get too rough, they pull up the drill so as not to damage the expensive equipment. Sometimes research ships need to return to a hole they sampled years before, and finding the hole again under 10,000 feet of water with a 1-foot-diameter drill in moderately calm seas requires an accurate location-finding system. A run of bad space weather can delay operations, which may not be financially feasible for the ship at sea.

Companies that operate pipelines (the Trans-Alaska Pipeline, which moves oil through Alaska, for example) watch the level of geomagnetic current (Fig. 7.10). Current in the pipeline can cause an electrochemical reaction that can corrode the huge steel pipes, causing leaks that are both expensive to mend and harmful to the environment. GICs flow best through conducting rock or ground, but as the steel pipeline conducts better than the ground, the currents travel through the pipeline. To protect the pipeline, electric current (moving opposite to the undesirable current) is induced into the pipeline to keep the total current flow as close to zero as possible. During large geomagnetic storms,

Figure 7.10: *The Trans-Alaska Pipeline transports oil from the oil fields near Prudhoe Bay on the Beaufort Sea in the north to the port of Valdez in the south. The long conductors, in the high latitudes, experience a lot of GICs. (NOAA)*

operators of the pipelines have to carefully monitor and balance that current. Like pipelines, the long, conducting lines of railroad tracks also suffer from geomagnetic activity. The GICs interfere with the small amount of power that runs along the rails and supplies electricity to the signals along the tracks. The current from a space weather storm can greatly exceed the railroad's electric current, potentially damaging components of the signal system not designed to handle such high levels. The failure of a signal can lead to expensive and tragic accidents.

People who race pigeons have learned the hard way not to hold competitions during even moderate geomagnetic storms (Fig. 7.11). The disruption confuses the pigeons, and they fail to return to their coops; as many as 90 percent of the birds can be lost in a bad space weather storm. The loss of an experienced racing pigeon can be devastating to the owner, and racing pigeons can be quite costly. Without space weather forecasts, the race could become what pigeon racers call a "smash." Pigeon racers have fought for permission to race birds in the vicinity of Chicago, whose city officials wished to remove all pigeons from the city for supposed health reasons. From the outcry this proposal generated, it is clear that several pigeon racing organizations and thousands of sporting enthusiasts support the races—and carefully watch space weather.

Some users even faithfully watch space weather forecasts in order to manage their investments in the stock market. Of course, for there to exist a scientifically believable link between space weather and the stock market, a mechanism for how they influence each other must be identified—and no such causal relationship exists. However, in recent years a correlative relationship has developed between space weather and the futures market. Power companies look at a large set of criteria when they establish the cost for energy. Californians, for one, know that the price and availability of energy can change minute by minute. The state's catastrophic energy supply-and-demand imbalance in 2001 caused great personal and state budgetary turmoil, not to mention political upheaval. Given the fluctuating demand for power and unusual supply changes, the "spot market" for energy can change greatly. Savvy investors watch space weather forecasts for activity that may lead to power outages or the lowering of the electricity supply to protect equipment. The threat of a storm will hail an increase in the cost of energy over the next one-day and three-day prices, thus making for lucrative futures market investments (Fig. 7.12).

Finally, a huge number of users are not negatively affected at all by geomagnetic storms but are, in fact, thrilled to hear of one. These people eagerly check the SEC website for forecasts of the aurora borealis and the aurora australis (Fig. 7.13). Even people living in the southern parts

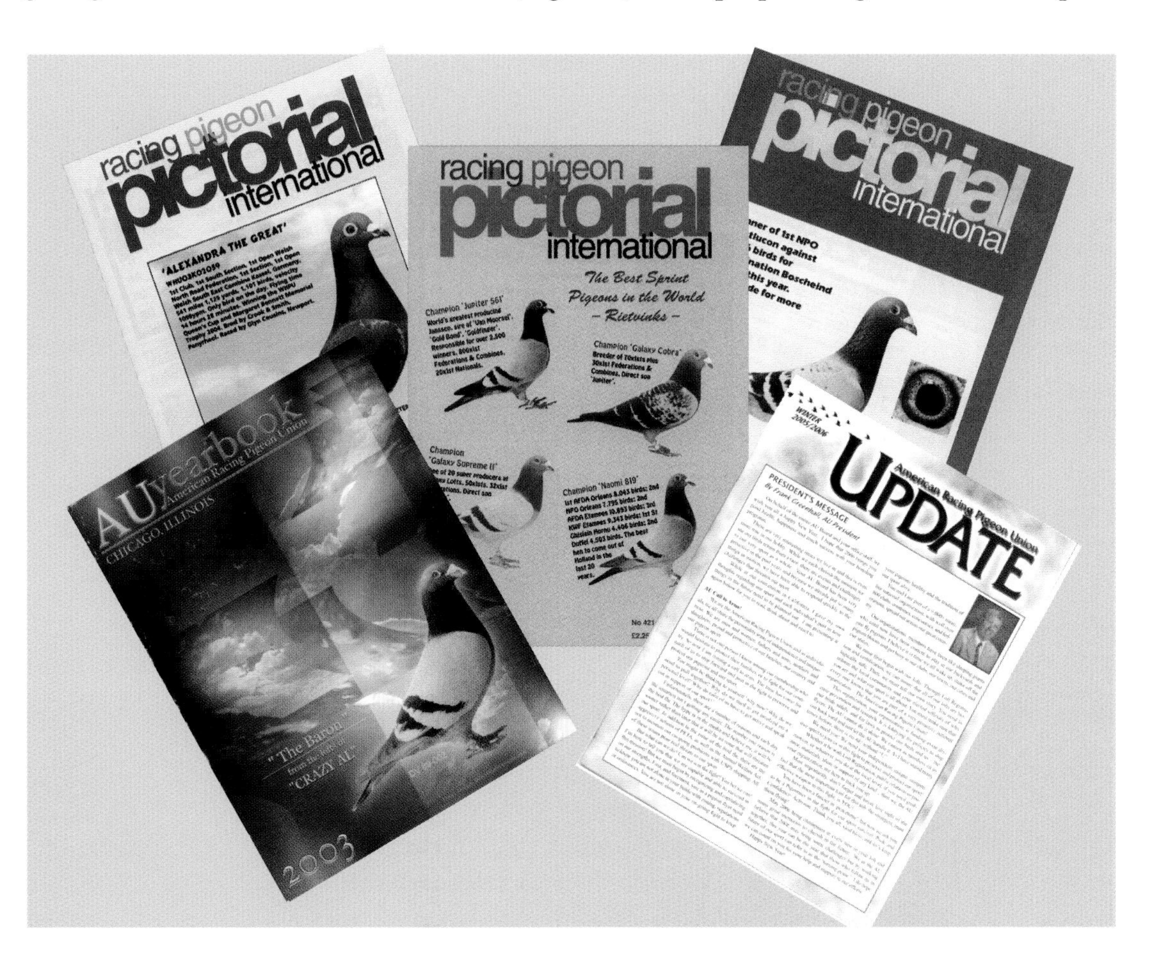

Figure 7.11: *Magazines and newsletters attest to the acute interest in the sport of pigeon racing. Several other navigating birds, mammals, and insects also get confused in a geomagnetic storm.*

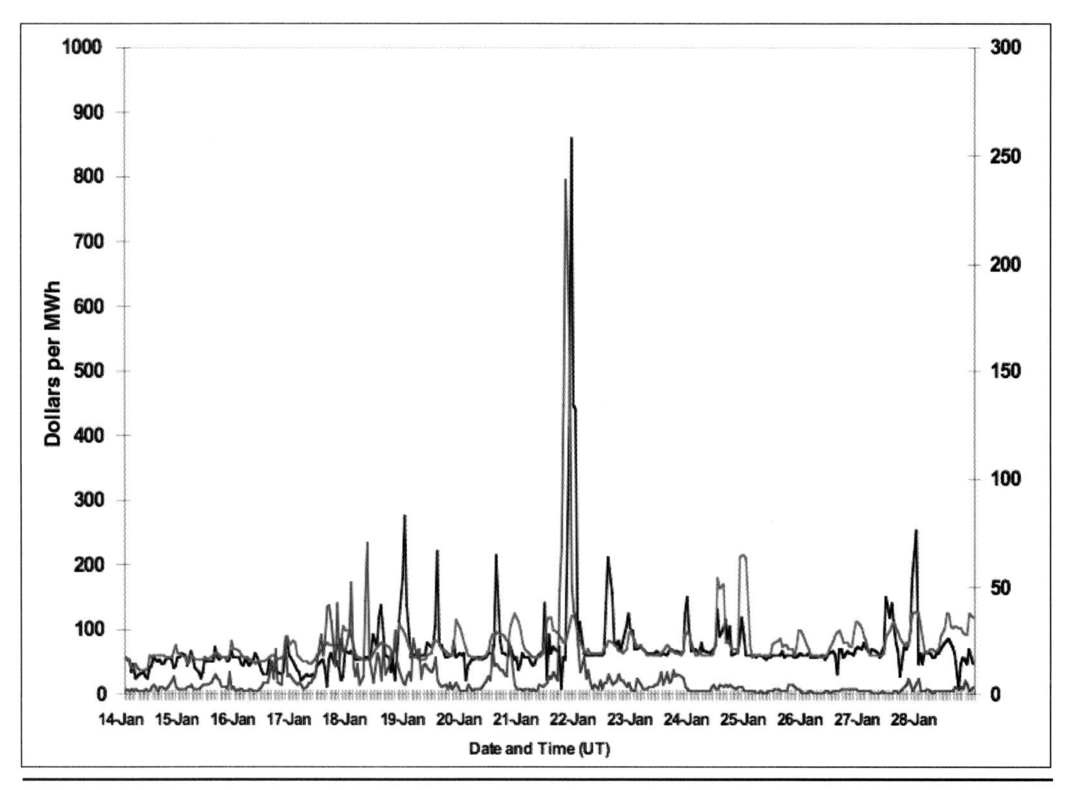

Figure 7.12: *The price of electricity on the New York energy market (in black) follows the rate of change of a geomagnetic disturbance (in red) on January 22–23, 2000. The day-ahead price is in blue. (Kevin Forbes, Catholic University of America)*

of the United States are able to see the lights during extreme storms. For many people, seeing the northern lights is a once-in-a-lifetime, unforgettable experience.

Solar Radiation Storms

Solar radiation storms increase the amount of highly energetic particles that come from the Sun and settle into the atmosphere of the Earth. Humans and satellite and high-frequency radio communications commonly have the most problem with solar radiation.

Aboard airplanes, certain electronics (along with passengers and crew) can be affected by radiation. Single-event upsets (SEUs, also caused by geomagnetic storms) interfere with electronics, causing what is akin to hitting a key on your computer that deletes your program or shuts off your machine. Reinstalling software or rebooting the system may repair damage done to electronics. Occasionally, however, the SEU destroys parts of the electronics that cannot be repaired, such as part of a memory chip. Flying has become more comfortable, safe, and easy because of the increase in electronic devices aboard the aircraft, but with that comes an increased vulnerability to SEUs. Although some electronics

Figure 7.13: *In these photos of the aurora australis, the white-blue aurora was seen near the south pole at the Antarctic Station, while the redder aurora was seen from a lower latitude in southern Australia. (NOAA)*

on airplanes (such as the entertainment system) provide merely comfort, the loss of which is only an inconvenience, electronics glitches in other parts of the plane can be serious. As airplanes increasingly use what is called "fly by wire," whereby the pilot controls the wing movement with wireless computer instructions, space weather SEUs can threaten the actual flying of the airplane. An accident resulting from this situation has not yet been reported, but the airlines are aware of this possibility and are trying to learn more about space weather to mitigate the risk. Most people worry about the radiation risk to humans in an airplane, but compared with the risk of a plane crash due to an SEU, the small long-term risk to an individual human body seems minor.

Scientists can be every bit as surprised as users to find out that space weather affects their systems. The National Institute of Standards and Technology (NIST) is located next door to SEC in Boulder and houses one of the few atomic clock standards. Atomic clocks are the most accurate clocks in existence, and the few scattered around the globe set the world time standard. A cesium atom, prodded by electromagnetic radiation at a specific frequency, vibrates between two states, creating a regular beat that the clock counts to register time. The NIST-7 cesium clock is accurate to five parts in 10^{15}, which means the clock gets off by one second every six million years. Space weather can cause that rock-solid ticker to miss a step because a solar magnetic field interferes with the magnetic fields generated by the vibration. Imagine the surprise of SEC staff when the scientists in the adjacent office said that they check space weather all the time to adjust the clock! SEC had no clue that this was a space weather problem.

Atomic clocks have uses aside from telling the time on Earth. Satellites send signals to ground receivers at the speed of light, and the time it takes between transmission and reception indicates the distance between the two. Atomic clocks' great time accuracy allows for greater accuracy in the measuring of distance and location. A little error in time amounts to a great error in distance, so time is crucial. Each of the twenty-four to twenty-nine GPS satellites contains four atomic clocks, which get an accurate time update daily from the U.S. Naval Observatory's atomic clock. The redundancy of clocks—two cesium and two rubidium—serves two purposes: if a problem arises specific to the cesium atom clocks, the rubidium clock ticks on unaffected; if one of each clock breaks, the other pair of clocks acts as a backup. The clocks have an average lifetime of about five years and the satellites are expected to run for three to seven years.

A group of scientists studying heavy water found they needed space weather information. In March 1989, scientists in Utah reported that fusion had occurred in experiments on the electrolysis of heavy water (D_2O). They did not detect heat (one indicator of a nuclear process) but did claim to have observed neutron emission, which would also indicate a nuclear process. The claims were particularly astounding given the simplicity of the equipment: just a pair of electrodes connected to a battery and immersed in a jar of D_2O—equipment easily available in many laboratories. (Humans have created fusion but only under enormous pressure and temperature.) Months later someone thought to call SEC to ask if space weather could have a bearing on the controversial findings. The scientists had measured an increase in neutrons and assumed it to be a product of fusion, but they had failed to consider that both cosmic rays and solar radiation could raise the neutron content in the atmosphere. SEC answered that yes, they did need to consider space weather and free neutrons.

Radio Blackouts

The final storm type, radio blackouts, is caused by the release of high-speed x-rays from solar flares. Radio blackouts, unlike geomagnetic and radiation storms, are neither carried by the solar wind nor affected by the Earth's magnetosphere; the x-rays that cause radio blackouts hit the day side of Earth at the same time that we see the flare. A large enough x-ray emission can cause a complete HF radio blackout on the entire sunlit side of the Earth, lasting for a number of hours, and can spread to the night side. A more moderate radio burst will disturb only an area of the sunlit side of Earth and last for less than an hour. Although geomagnetic storms affect HF radio communications primarily at higher latitudes, radio blackouts affect systems and users at all latitudes.

As mentioned several times before, mariners and general aviators using HF communications risk having no radio contact during a blackout. Luckily, extreme blackouts happen only a few times in a solar cycle (every eleven years). The ionosphere exists because of x-rays from the Sun that have ionized particles that are in that region of the atmosphere. Radio blackouts are merely increased x-ray activity, which further ionizes the ionosphere; as mentioned previously, radio signals travel unpredictably through thicker regions of charged particles. As a result, low-frequency navigation (such as Loran) used by maritime and aviation systems loses radio contact for perhaps an hour during a moderate blackout or for many hours in an extreme blackout. Satellite navigation can also be plagued by errors in positioning.

Cell-phone transmitter towers are not usually affected by radio blackouts, but problems can occur in specific circumstances. When a cell phone sends a signal to its communication tower, or between towers, and the signal is directed in line with the Sun, solar x-rays can wipe out the signal. It is analogous to trying to drive with brilliant sunshine on the windshield: the bright x-rays simply swamp the signal. Although this happens only about 2 percent of the time, the occurrence seems significant when the number of cell-phone users is so high.

Along with the great variety of weather and space weather instruments, the NOAA Geostationary Operational Environmental Satellite carries a small radio transmitter and receiver that can pick up beacon signals from distressed individuals at sea or on land and pass along the message to NOAA. The NOAA satellite is always listening for these distress signals and can relay an SOS very quickly. Unfortunately, a troubled individual may not be able to send the signal for very long, sometimes only for a minute or two. During a radio blackout, any SOS will fall on deaf ears, as the signal cannot get through to the satellite. NOAA would never know about an emergency broadcast for three minutes during a ten-minute outage. Such an outage occurred during a large flare in November 2004. The blackout lasted for one hundred minutes and was a system failure that could have cost lives.

The number of users of space weather services has increased dramatically in the last sixty years, largely due to the striking growth of technologies. Perhaps surprising is the number of technologies that have not engineered their way out of space weather problems. Either the affected technology is inherently trapped by space weather effects (such as communications that require the ionosphere to work, and the ionosphere cannot be modified by humans), or competing needs negate the solution (such as the high cost of engineering around the risks).

So, if users cannot avoid space weather, what can they do? Although the answer largely depends on the type of storm and technology, mitigation is far more likely than overcoming damage. The forecast tells how big a storm will be, when it will arrive, how long it will last, and where it will be most damaging. Given this information, users must consider their specific technology, as small storms do not necessarily cause little damage and big ones more; it depends on the technology. Satellites outside the magnetosphere, for instance, can be vulnerable to small disturbances, especially long-duration ones. Valid predictions allow the user to choose what to do. But some users want more than just the forecast. Some want to see a certain percentage of correct predictions before they will even use the forecast; some want past warnings for retrospective study; some just want to have the data for their own interpretation. Many want new space weather models, new data, or new products. Some of these fundamental user needs came up at the Solar Prediction Workshops, international meetings held in 1979, 1984, 1989, 1992, and 1996. Although these workshops pulled in users and researchers, they were small and fairly technical. Users wanted a broader voice in describing their needs, and they began to get that direct interaction with SEC and other space weather forecasters and researchers in 1990.

In 1990, the Space Environment Center had 2,500 steady customers who had signed up to receive weekly reports on space weather. Although some were solar physicists and researchers, many were operators of technologies who may have taken preventive measures against space weather events but who understood little of the purpose behind those measures. They were not necessarily trained in the science of space weather, and while they knew how to enter a "K7" (an index of geomagnetic disturbance) in their logbooks, they did not really know what that number meant. Users needed more. Three major additions have helped the space weather community since 1990.

The first addition began in 1990—a meeting of users held in Boulder, the predecessor to Space Weather Week. Although the meeting attracted only a few dozen users, many SEC staff members and a few researchers attended, all curious to find out what users needed. Various breakout sessions on specific areas (HF communications, induced currents, navigation, satellites) promoted open, relaxed dialogue. Many of the users were self-conscious about their limited understanding of the field, and were shy about speaking up. Nevertheless, many gave SEC useful guidance. Most of all, SEC realized how much training it owed to the users.

SEC also ran a yearly series of research meetings starting in 1995, separate from the user meetings. Eventually, the logic of having a joint meeting with both users and researchers became obvious. Space Weather Week began in 1998 as a three-and-a-half-day meeting, attracting members of the new National Space Weather Program; organizations and individuals who provide funding; researchers from international government agencies, academic institutions, and private companies; operators; users; and forecasters. The meeting proved to be a great success and has grown and evolved into a unique and important conference (Fig. 7.14). It is the only meeting of its kind, so far, to successfully balance the users with forecasters and researchers in the field. The issues introduced and discussed at Space Weather Week set scientific direction, state vital user

Figure 7.14: *Space Weather Week is all about people learning about and dealing with space weather. The annual meeting draws those most interested in understanding the impacts of space weather storms and what they can do to help their technologies avoid damage and failure. (NOAA)*

needs, beget changes in space weather services, and draw the space weather community into a close network.

The second major addition to the space weather community came with the life-changing Internet, which has flung open the doors of space weather research and forecasting to the everyday user. Although the government began using the Internet for e-mail and file transfer back in the 1970s, most users had no access, so the usefulness of the early Internet was limited. But as soon as the Web became useful to civilians, even when it contained only a few hundred sites, SEC had a website that posted daily products and forecasts. Now virtually all data and products are online and accessible within seconds or minutes of SEC receiving them. Perhaps best of all, the Internet has opened the doors to interested nonusers—teachers, students, professionals, and those simply intrigued by space weather. Although SEC has thousands of users, its website's daily hit rate numbers in the hundreds of thousands. As space weather grows in people's awareness, so do the hits, especially during periods of solar activity. And when the hit rate grows, it never falls back to preactivity levels. The October–November 2003 storms registered an all-time high of nineteen million hits in twenty-four hours. After the storms, levels dropped back to slightly above-normal levels of 600,000 hits a day. E-mail has greatly increased the ease of delivery of products and forecasts while reducing the costs that come with postal mail.

The final addition to space weather services was the commercial vendors. Starting in 1995, SEC chose to use the guidelines and policies of the National Weather Service (NWS) regarding commercial vendors. Many years ago NWS had run aground with Congress when some commercial groups protested what they regarded as the government's denial of public competition. According to these groups, NWS had crossed the line that separated what the government agency could provide for free and what commercial vendors could sell. The controversy started because in 1954, NOAA forecasters regularly appeared on TV to deliver the terrestrial weather prediction. This simple action blew up in their faces. The American Federation of Television and Radio Artists filed a complaint against the NWS requesting the prohibited non–union members from appearing on TV. Congress sharply directed NWS to develop and support a commercial weather industry, not compete with vendors for TV airtime. Congress thought it desirable to keep the ability (and liability) of forecasting weather under government control but allow commercial companies to present the NWS forecasts to the public. Today virtually every newspaper, news program, and TV station such as the Weather Channel shows evidence of commercialization. A natural tension exists between vendors and organizations such as NWS because the vendors see NWS as an advantaged competitor, while NWS sees itself as giving the taxpayers what they pay for. A careful debate is ongoing and will never be truly resolved because both sides feel their obligations keenly.

Despite the possibility for tension, SEC entered partnerships with vendors and got along with them fairly well. SEC had always gotten along very well with its customers, some of whom were toying with the idea of starting a company to sell space weather services. Following the NWS policies, SEC accepted a Cooperative Research and Development Agreement (CRADA) with Sterling Software in 1996 for the purpose of developing a magnetospheric specification model. Both SEC and Sterling Software benefited from this interaction. A second CRADA with Space Environment Technologies started in 2001 and has yielded an ambitious set of model products used by both SEC and paying customers. SEC has also supported some Small Business Innovative Research grants, one of which will develop a new geomagnetic activity model for use by power companies.

Other companies are eager to start providing space weather services, but getting off the ground has proven to be difficult. One good sign of progress is the formation of a vendor group, the Commercial Space Weather Interest Group. Vendors in the group range from those who sell space weather products to power companies and to the USAF, which supports GPS and HF communications. The dozen or so members work together and individually to solicit help or information from SEC and have lobbied in favor of supporting SEC so that they can benefit from the government work. At this point, the space weather industry is a long way from making the kind of money that commercial weather companies make, but it continues to try.

Ultimately, the bottom line for quantifying the value of space weather lies with the users. They operate the infrastructure that keeps the economy flourishing and maintains our way of life. The socioeconomic effects of space weather are evident in all direct losses (delays and damage) and prevention costs suffered by these users. Many people in the field have tried to add up the commercial costs of space weather, but the total fails to take into account several factors, including the incalculable personal costs associated with disruptions and failures. Despite the value to the military, satellite communications systems, navigation systems, power networks, astronauts, high-altitude aviators, scientists, and all other users, SEC often must defend the work it does to the very organization that created it—the government. To tease apart that puzzle requires a look toward Washington, D.C., not the Sun.

8 Ironies of Attention and Neglect

In any large bureaucracy, the distance between the highest and lowest branches can turn into chasms of misunderstanding. The highest branches may envision a world where everyone uses paper clips and thus send truckloads of paper clips to the lowest branch, which specializes in the use of staplers. Throughout every reorganization of the Space Environment Center (SEC), from a quasi-military group in the Central Radio Propagation Laboratory (CRPL) to a service group in the National Weather Service (NWS) of the National Oceanic and Atmospheric Administration (NOAA), the organization's emphasis has often been decided by someone at upper levels in the government—the Department of Commerce, Congress, or presidential advisers—rather than by the scientists and forecasters who understand the needs of the users. Sometimes these bureaucratic decisions were excellent; others were ill-considered. Over the years the status of space weather has swung from attention to neglect, from favor to disfavor, and back again.

As discussed in Chapter 3, the history of any governmental organization can be extremely complex. But the renaming, reorganization, and refocusing of agencies are typical of how the government works in the United States. A more detailed explanation of the history of SEC will give a good impression of the relationship between the science of space weather and the politics of space weather (Fig. 8.1).

CRPL (1945), the earliest U.S. space weather–ionospheric research organization, fit well into its founding agency, the National Bureau of Standards (NBS). NBS's responsibility lay in issuing official time, setting industrial measures, and developing commercial applications such as closed captioning for television. CRPL focused on the ionosphere, which had a standards and regulatory aspect; for instance, commercial broadcasting needed to be assigned specific frequencies at which to transmit.

In 1963 Dr. J. Herbert Hollomon, assistant secretary of commerce for science and technology, decided to merge several agencies within the superagency of the Department of Commerce (DoC): the Weather Bureau, the Coast and Geodetic Survey, and CRPL (of the NBS). The resulting organization, called the Environmental Science Service Agency (ESSA), officially started in 1965. It was rumored that Hollomon considered the Weather Bureau and the Coast and Geodetic Survey to be two

The Family Tree of the Space Environment Center
(All Within the Department of Commerce)

Year	Organization
	NBS National Bureau of Standards
1941	IRPL Interservice Radio Propagation Laboratory at Ft. Belvoir and in D.C. locations
1945	CRPL Central Radio Propagation Laboratory
	SEFD Space Environment Forecasting Division
1965	ESSA Environmental Science Service
	SDL Space Disturbances Laboratory (formerly CRPL)
	SDFC Space Disturbances Forecast Center
1970	NOAA National Oceanic and Atmospheric Administration
	ERL Environmental Research Laboratories
	OAR Office of Oceanic and Atmospheric Research
	SEL Space Environment Laboratory (previously SDL)
	SESC Space Environment Services Center (previously SDFC)
1995	OAR Office of Oceanic and Atmospheric Research jointly under NWS/NCEP
	SEC Space Environment Center (previously SEL)
	SWO Environment Services Center (previously SESC)
2005	NWS National Weather Service
	NCEP National Center for Environmental Prediction
	SEC Space Environment Center
	FAB Forecast and Analysis Branch and SEC Forecast Center (previously SWO)

Figure 8.1: *The family tree of the Space Environment Center. (NOAA)*

old-line service organizations with relatively weak research components, and that these groups could benefit from the well-regarded CRPL, a five-hundred-person research institute. The transfer of CRPL from an agency focused on standards to one focused on environmental science such as measuring physical parameters and tracking environmental changes broadened the research opportunities for CRPL scientists and environmental science engineers. ESSA had both operational responsibilities and a strong research component. This cobbled-together agency, according to Director Gordon Little, was "a unique program in the Nation, because it covers within one administrative unit a strong integrated program of theoretical, laboratory, ground-based, and rocket and satellite studies of the upper atmosphere."

In 1964 a government committee determined that the DoC and, by proxy, its various suborganizations would be given the central federal responsibility for "space weather forecasting." The National Aeronautics and Space Administration (NASA), the Department of Defense (DoD), and the Federal Aviation Agency (FAA) eagerly awaited the external support this assignment would provide. They all had great interest in space weather but did not wish to start forecasting in their own agencies. So in February 1965, in response to the assignment, the ionosphere research and propagation part of CRPL was reorganized into the Space Disturbances Laboratory (SDL), which included a forecasting center, the Space Disturbances Forecast Center (SDFC). SDFC issued the first of its daily Space Disturbances Forecasts in 1965 to support the Gemini flights.

In 1970 several of ESSA's divisions were moved to the National Bureau of Standards, and others moved to the newly formed National Oceanic and Atmospheric Administration (Fig. 8.2). SDL was renamed Space Environment Laboratory (SEL) and made one of the Environmental Research Laboratories in NOAA. Here the schism mentioned in Chapter 3 occurred between SEL and the Space Environment Services Center (SESC). Don Williams took over SEL and steered the research group toward the area that was his passion—the magnetosphere (especially on Jupiter), not the ionosphere. Funding of ionospheric research had flagged markedly at this time because of the successful strides that had occurred since World War II. By the 1970s interest in the iono-

Figure 8.2: *The Hoover Building houses the Department of Commerce, and within it NOAA's headquarters. The building sits along the Mall in Washington, D.C. (U.S. Department of Commerce, Photographic Services OS/OPA)*

sphere had shrunk to such an extent that there were massive layoffs in, or attrition from, the ionospheric-science field. Most of the people funding space weather thought that enough had been learned about the ionosphere and that satellites would displace the need for technology using the ionosphere. They, like Williams, were eager to reach into space and explore where satellites flew.

Bob Doeker, head of the newly named SESC (previously SDFC), held just as passionate views as Don Williams. Even though satellite communication had overtaken the reliance on high-frequency (HF) communications, and the rest of the community had shifted its focus elsewhere, the U.S. Air Force (USAF) persisted in its support of HF communications. Doeker wanted to involve the USAF in SDFC operations. He pursued strong monetary support from the Air Force and began the very tight relationship between the two agencies that ultimately led to a few Air Force personnel being stationed at SESC in 1972. In this partnership SESC provided solar forecasts to the Air Force, which focused on the effects in the ionosphere and geomagnetic predictions. In this manner the USAF received the solar observations and forecasts it needed and the SESC received the USAF products that allowed access for civilian users to military information. Doeker played an important role in developing the nation's real-time space weather services through the reorganization.

The split between forecasting and research hurt the productivity and efficiency of the organizations. The three major elements of space weather services—research, forecasts, and user needs—have long existed in a symbiotic relationship in which each element drives the others. New technology from researchers allows for better forecasts from forecasters, which better serves users, who in turn express their demands for improved technology. Recognizing the need for healthy cooperation, Ernie Hildner, the director of SEC (previously SEL), beginning in 1986, worked hard to bring the two long-separated halves of SEC together. Because of his persistence, the efforts of his deputy, Ron Zwickl, and the support of other staff members, SEC has become a model of collaboration between research and operations. The strong mutual support of the two branches led to a shift toward the National Centers for Environmental Prediction (part of the NWS) in 1995 and a complete move to the NWS in 2005.

In the forming of NOAA back in 1970, the SEL had been put into the federal budget as a line item. That meant that members of Congress and their staffers could plainly see what funding was going to this particular program. It also meant that Congress could target that specific program for more or less funding. In a way, this approach resembles high-stakes gambling in Las Vegas. By being a line item in the congressional budget, a program's funding, favorable or not, is set by law. Of course the alternative, being lost in the NOAA budget (not a congressional line item), means a program can be favored or cut in the discretionary internal budget—not exactly a stable funding situation either. The SEC, with its small annual need of $4 million (a number that has grown over the years), crouched among the big agencies and major projects that needed massive funding. There, it was easy to fund or easy to ignore.

Being in such a precarious funding position, SEC inevitably faced numerous budget crises throughout the years. To understand these crises, it is helpful to walk through the process of government funding. The procedure can be described as a bottom-up, then top-down process. First SEC (at the bottom of the hierarchy) identifies its funding needs. For the fiscal year 2007 (FY07; for the government this is October 1, 2006, to September 30, 2007) budget, still two years away, an e-mail will arrive from NOAA headquarters asking for SEC's projected spending needs, required to be sent back to headquarters by the end of the day. Should the next level up the chain approve the desired funding, celebration can last only until the next layer up—or the next, or the next—also approves it. Headquarters (in Washington, D.C.) can accept the budget as it is or modify whatever projection SEC sends. Headquarters then forwards the collective NOAA budget up the chain, through the NOAA administration to DoC, and then to the Office of Management and Budget, from which the presidential budget request is crafted. When Congress receives the presidential budget request, it sends it to the House of Representatives and Senate appropriations committees. When the considered budget comes out of committee, each branch of Congress debates it in full and sends the two approved versions to a joint appropriations committee to reconcile the differences. The compromise version can look very different from what the House and Senate sent to the joint committee. Congress takes the committee's compromise appropriations bill and votes on the entire thirty-thousand-page document after what may be only a few hours. Whatever the full House and Senate vote on is final, whether the decision harms or benefits SEC. This is the bottom-up aspect of the process.

The top-down part begins with the approved federal budget. The money can trickle down to the agency in two ways, depending on whether the project holds line-item status. For non-line-item projects, the agency that received the funding chooses how to divide up the money among its subagencies—a process that has actually been pretty well worked out in advance. In other words, if Congress gives NOAA a pot of money, NOAA is responsible, for the most part, for choosing how to spend it. If NOAA receives less money than it requested, its managers must make the hard decisions about what to cut back. Congress can still put restrictions on the funding and say that NOAA has such-and-such money but cannot spend it on administration, or that it must spend it on climate research. As SEC is a line item in the budget, NOAA is constrained to give SEC the amount of money Congress allocates for it. That means that no matter what the size of the NOAA budget is, SEC will get its money. SEC has not always been a line item and no longer will be once it is strictly under NWS funding.

Most members of Congress have limited knowledge of each aspect of the budget they must consider. SEC values like gold the few members of Congress who make an effort to learn about space weather. Congress can request information about space weather, and congressional staffers can be invited to visit SEC, but SEC must be a passive host. Legally constrained from "lobbying" congressional members (and restricted from even speaking to them without "handlers"), the government employees can only hope that Congress has enough interest in space weather to ask for information. In some cases the line between being a passive host and lobbying grows very thin. Occasionally an employee chooses to walk that fine line, which leads sometimes to censure and sometimes to success—a tricky risk for the employee.

Obviously, the budget does not always get approved, and SEC has faced various financial crises over the years. Funding for SDL in the 1970s was tight, but circumstances required that the lab be funded. The NASA missions, spurred on by Cold War fears, required an abundance of staff and products from SEL. USAF staff helped out at SEL, and NOAA corps officers joined SEC staff at NASA to manage the high demand; NASA provided much of the extra money needed.

The first budget crisis hit SEL in 1983, the result of three shaping events. First, a severe "irregularity" occurred in the laboratory's financing, which SEL management wanted to keep quiet. In order to do this, Don Williams, director of SEL, had to balance his budget by reducing promised finances to the various projects in the lab. Second, Williams's scientific interests involved the study of the magnetosphere on Jupiter. Magnetospheric physics was not specifically in the NOAA mission, but as long as the studies involved Earth, NOAA found them acceptable. Jupiter, however, lay well beyond the mission of NOAA, and this research exhausted NOAA's patience with the lab. Third, during his administration President Ronald Reagan worked hard to reduce the size of the government. In fact, Reagan planned to cut taxes while spending money on somewhat secretive military operations and needed executive branch funding for these other projects. Congress went searching for "unnecessary" funds, and space weather must have looked unnecessary.

When Don Williams left the lab in March 1982 to pursue a vital scientific career at Johns Hopkins University, Harold Leinbach took over the troubled SEL as acting director. Glenn Jean, a leader in the lab since the 1960s, acted as Leinbach's assistant director. The funding situation was precarious. Although SEL was a line item in the president's budget request, NOAA did not fully support SEL, and it retained the right to modify SEL's original budget requests. Leinbach recalled, "At the time I took over the lab, we were under the gun for a huge budget cut and redirection of the lab's efforts with deemphasis on research not directly pertinent to NOAA goals." Leinbach, Jean, and others in the lab spent days in Washington, D.C., talking to NOAA and DoC administrators and, to the extent that they could, to congressional staff. Congress considered zeroing out the SEL budget. Luckily, backing SEL was a powerful group unwilling to see space weather services eliminated—the users.

As Congress worked toward a consensus on the budget, space weather service users aided in finding a resolution. Unlike federal employees, these constituents were free to talk to members of Congress, and constituents and lobbyists have a remarkable talent for getting the ear of a legislator. The users, the voting public, had experienced how valuable SEL could be to their interests, and they informed the members of Congress that space weather should have funding priority. At one point during the crisis, SEL stopped sending the hourly message about HF radio problems as a way of cutting unfunded activities. It received a quick response from Senator Barry Goldwater (R–AZ), who howled in protest; he was a ham (HF radio operator), and he wanted that space weather information.

Senator Tim Wirth (D–CO), with a little luck, saved the day. He sat on the finance committee and had agreed to push for fully funding SEL. But, as Wirth's staffer reminded him toward the end of the meeting, the committee (and Wirth himself) had, in fact, omitted discussion of the small matter of SEL funding. With the budget process virtually over, there could be no more votes. Wirth, in a private conversation with the committee chairman, admitted this lapse. The chairman agreed to have Wirth ask about the line item in a procedural meeting. In that meeting, the chairman responded to Wirth's question, noting that the SEL budget had been inadvertently omitted in the final version, which would be amended. A procedural trick at the last moment saved the entire program of space weather in NOAA.

The importance of space weather to the government increased in 1995, when a broad-based group of high-level agencies formed a new program, the National Space Weather Program (NSWP). NSWP became a multiagency federal research program seeking to mitigate the adverse effects of space weather. The NSWP Strategic Plan describes its goal: "The overarching goal of the National Space Weather Program is to achieve, within the next 10 years, an active, synergistic, interagency system to provide timely, accurate, and reliable space environment observations, specifications, and forecasts. It will build on existing capabilities and establish an aggressive, coordinated process to set national priorities, focus agency efforts, and leverage resources. The Program includes contributions from the user community, operational forecasters, researchers, modelers, and experts in instruments, communications, and data processing and analysis."

NSWP participants are the Department of Defense, Department of Commerce, Department of the Interior, Department of Energy, National Science Foundation, NASA, and the Office of the Federal Coordinating Meteorologist. The coordinated effort of such a broad reach of agencies has resulted in excellent research in space weather forecasting. The interagency program helps maintain the network and communication among everyone working in the field of the solar-terrestrial environment. The largest contributors are NASA and the National Science Foundation, funding excellent research projects; the Department of Defense, contracting for specific research development; and the Department of Commerce—SEC. Space Weather Week plays a key part in evaluating the progress made from year to year.

Despite the formation of NSWP in 1995, SEC limped through the following decade, during which the funding levels did not keep up with inflation. The president can dictate raises for government employees (including cost-of-living raises) in his budget proposal, but Congress may or may not choose to fund them. The agencies then have to find money for the raises in their own limited budgets. During the years after 1995, Congress did not increase funds enough to cover these raises, leaving SEC with obligations to its employees and programs that it could not meet. Luckily, a string of retirements lowered the employee count and kept the funding in line; unfortunately, the reduction in staff did nothing to help with the increasing demands on the center. Bill Clinton's presidency saw a slow and insistent reduction in the number of government employees. At one point NOAA prepared to institute a reduction in force, the government's way of laying off staff, a painful acceptance of the lack of funding. This plan was never carried out, as NOAA had no stomach for it; instead, it chose to squeeze dollars out of nonlabor funds.

Then began another push to reduce government during George W. Bush's administration. For the FY03 budget, suddenly SEC, which had been "flat funded" (successively more poorly funded) for many years, found itself on the block for elimination. Indeed, all of NOAA faced termination. Debates in the House addressed various ways to reduce the size of organizations and their number of tasks, especially if the private sector could do the work, with or without contract funds.

A Cato Institute report, prepared by Edward L. Hudgins, summarized with startling hyperbole the state of some congressional feeling in 1999: "Congress should close the Department of Commerce and, in particular ... phase out the functions of the National Oceanic and Atmospheric Administration, allowing them to be supplied by the private sector. ... Most of NOAA's activities can be eliminated immediately and associated assets and equipment sold off to private bidders. Since weather monitoring and forecasting can be performed by private-sector providers, the National Weather Service should also be privatized. A transition plan might be necessary to back the government out of the weather business."

This summary reflected the sentiment of Representative Dick Chrysler (R–MI), who explained why he had voted to eliminate the NWS during the Republican Congress's first one hundred days in 1995—"We don't need the government; I get my weather from the Weather Channel." This was a rueful joke to everyone who actually watches the Weather Channel and sees the repeated credits to the National Weather Service during every broadcast. "I get all the news I need on the weather report," sang Simon and Garfunkel, and those reports would not have existed without the NWS's satellites, data, computers, expert meteorologists, and product dissemination.

The National Academy of Sciences weighed in on the side of the agencies on January 31, 2003, with a panel finding that the NWS was vital. A report in the Pittsburgh, Pennsylvania, paper, the *Post-Gazette*, by staff writer Don Hopey explained:

> Although the political climate favors private enterprise, you still need a National Weather Service to tell which way the wind blows and do other data collection and forecasting chores, according to a new report from the National Academy of Sciences.
>
> The report issued yesterday says the Weather Service should continue to issue daily forecasts and weather advisories, rebuffing a storm of complaints from private meteorology and forecasting companies who say the free weather information the government provides is competing with services they can sell.

Some congressional staffers, it seemed, just plain did not like NOAA, NWS, or SEC. Some spread unfavorable comments around Washington about visits to the lab and various programs. One sore point that earned the enmity of one staffer was the building of a new NOAA building in Boulder.

The building for the Colorado branch of NBS was located on the DoC campus on Broadway Street in Boulder, a large plot of land given to the DoC by the Boulder Chamber of Commerce in the early 1950s. The chamber had purchased two adjacent ranches, which it offered to DoC because it was interested in the benefits federal commerce would bring to the city. The city of Boulder received $1 from the DoC in payment for the site. The NBS (now named the National Institute of Standards and Technology [NIST]) facility, also housing some of NOAA, occupied a small section of the total property. In the reorganizations in the mid-1960s, the environmental laboratories of DoC spread out to three buildings across Boulder, making collaboration among the various NOAA labs difficult. To make it worse, by 1988 one of the buildings did not meet fire safety standards, and the labs faced eviction. The University of Colorado, which owned the space, wanted it back. NOAA started seriously proposing the construction of a tailor-made building somewhere on the large DoC campus.

Thus began, in 1989, the effort to build a NOAA building that would house all the local NOAA employees. Those who did not want any more federal development on the Broadway property suggested several alternative locations, but the price of land drew the planners back to the free federal site next to the NIST building. The campus had plenty of space for a substantial addition, and the new building could benefit from its proximity to NIST. Planners envisioned sharing some of NIST's facilities, saving money by avoiding the need to build a library or a cafeteria. In reality, the plan had several flaws. The NIST building housed about one thousand people; the NOAA building would do the same. Neither the cafeteria nor the science library in the NIST building would be adequate for both organizations.

When given a complete plan, the city of Boulder balked. Boulder had led the nation in buying open space and limiting growth, even commercial expansion. Residents prized the lifestyle created by the abundance of open space and building restrictions. Boulder residents had found the open land surrounding the NIST building attractive for hikes on the trails that had sprung up there, and for access to other open space to the west. Neighbors of the property would not accept a disruption of their mountain views. They did not want development, least of all a large federal facility, any closer to them. They frowned upon having twice the number of federal employees, two thousand, streaming through their neighborhoods to get to work each morning and insisted that access to the facility, already limited after the bombing of the federal building in Oklahoma City, continue to be limited to a single entrance. Hikers and dog walkers wanted to enjoy what they had long regarded as an open park space, not government land.

It came as no surprise, then, that neighbors rejoiced when "evidence" surfaced that the site was an old Native American burial site, later claimed to be part of a medicine wheel. The stones said to mark the burial site or the center of the medicine wheel, however, turned out to be made of slag (a remnant of smelting ore). Furthermore, the stones had been dug up and thrown around during the fifty years of government ownership, so the resulting "ring" could not have been an original medicine wheel. Whatever the facts of the wheel, government policies for major construction required an Environmental Impact Statement, including, in this case, consultation with Native American tribes. Three meetings were held with leaders from fourteen regional tribes and a group of Native Americans concerned with medicine wheel heritage sites. The tribes agreed that the site was indeed sacred because their nomadic ancestors had used it for camping, hunting, and ceremonies, but they came to no agreement about the past existence of a medicine wheel. In the end NOAA signed an agreement with eleven of the tribes saying that the building would not stand on the sacred ground but would be built on another section of the DoC campus.

Finally, a working irrigation ditch ran across the property, and in the West, where water rights are inviolate, the ditch could not be moved. NOAA nevertheless found that building on the existing campus made the most sense and that ultimately the property did belong to the government. Local members of Congress worked on both sides. They urged the DoC not to flex its muscles and build regardless of the city's protests, and they chided the city for being so rigid, since the property did belong to the government. Finally everyone accepted the site (Fig. 8.3).

Initially, according to the City Council, the blueprints presented a building too high, too big, and in the wrong place. One architectural firm was fired, and the planning started over. NOAA scientists carefully spelled out their needs but failed to consider the common needs in the building: a library, a classroom, a large and functional cafeteria, and a sufficient lobby. A small Common Spaces Committee was formed to fight for these areas, since the space for the common needs would have to be taken away from laboratory space. SEC director Ernie Hildner donated some of SEC's space for a library and the NOAA store, as SEC's need for space had diminished due to low funding. Through the committee, the building acquired a little art, display boards for science posters, a small branch library, a nice cafeteria, a classroom, a fitness center, and a half-sized model of the Geostationary Operational Environmental Satellite (Fig. 8.4). A new architectural firm accommodated both the NOAA tenants and the city and presented drawings proving that the building would not block the mountain view (Fig. 8.5).

Still, the compromises did not satisfy all concerned parties. With frustration mounting on both sides, U.S. Representative David Skaggs stepped in. NOAA and the city of Boulder painfully negotiated the configuration of the building—its site position, length, height, parking,

Figure 8.3: *The NIST site (right-most building) in Boulder, Colorado, opened in 1954, and the NOAA building (on the left) opened in 1999. NOAA and the city agreed to set aside some open space between the NOAA building and Broadway Street (bottom). (Colin Farrell Photography, Inc., © 2005, www.cfpix.com)*

access to the roof, and paved pads for research trailers and satellite dishes—with the patient urging of Representative Skaggs (Fig. 8.6). The project faced capricious funding from the government. Funding was available, then rescinded, then restored, and then threatened to be withdrawn. Skaggs, in the end, obtained the proper amount of funding and an accord between the agency and the city. In recognition of these efforts, congressional leaders elected to name the new building the David Skaggs Research Center (DSRC). Construction began in November 1996, and after a few delays and extremely cold weather, the building opened in early February 1999. The new tenants settled into a work environment with raised floors, allowing easily strung electrical and network cables, which can provide access points anywhere in a room without unsightly and unsafe cables running everywhere. The exterior offices were small but all had beautiful views. The building was energy-efficient and used local material that blended with the surroundings.

The U.S. General Services Administration (GSA) actually built the DSRC building for NOAA and rents it to the agency yearly. To make the building maximally useful for high-tech, world-class research and

Figure 8.4: *The sculpture in the lobby of the DSRC building depicts all aspects of NOAA science in the building—weather, the poles, data management, space weather, and satellites. The artists worked with the Common Spaces Committee on the design, and the committee funded the work. The sculpture was designed by Cathy McNeil and constructed by Dan McNeil and Cathy McNeil in 2001. (Herb Sauer, NOAA)*

operations, NOAA pitched in another 15 percent on top of the total cost to make the "above-standard" tenant improvements that GSA was not prepared to fund. It is a beautiful facility, carved out of innumerable compromises, brilliantly designed by many, and a monument to perseverance. It came with one major downside: the rent. All of the NOAA labs in the new building had to pay a higher rent out of their program funds, even though they had had no choice in moving to the new building. Acquiring funding for the building's additional rent expense has occupied the various lab directors since the building's inauguration. In the past few years Congress has appropriated a rent supplement to NOAA on a year-by-year basis, which has certainly helped. However, this is not a long-range solution.

The beauty of the new building perhaps caused a congressional staffer who visited it in 2002 to resent NOAA. The facility did not look like the typical block-type government building. Presumably too much money had been spent on this excessively nice facility. Shocked that visitors had deemed the building "too nice," the tenants were not ready

Figure 8.5: *The natural materials and the trees that happily grow near the irrigation ditch in front of the building made for a lovely site. The Flatirons remain in view from Broadway, as the city insisted. (Will von Dauster, NOAA)*

to defend it. They might have said that a nice facility attracts the excellent scientists NOAA wants and that the building results in high productivity and worker satisfaction. NOAA had needed that facility, and there it was.

The FY03 appropriation, passed in February (five months after the start of the fiscal year), allotted a shockingly small amount for SEC. The president's proposed budget and both House and Senate versions of the budget bill all agreed to fund SEC at the FY02 amount. However, somehow the joint bill that passed into law gave SEC only 60 percent of that funding. This outcome stunned SEC staff and could not be explained. Most distressing was that the announcement came more than four and a half months into the fiscal year, during which time government agencies had been told to spend at the previous year's rate. Somehow SEC had to make ends meet. Even if it planned to lay off staff (the only possible source of large "savings" in its budget), it was already too late; given government-mandated procedures and notification requirements, SEC could not get the staff off the payroll fast enough to balance the budget. Of course, the last thing SEC wanted was to abandon customers and research projects and terminate valuable staff. In the end, NOAA did not even authorize a reduction in force, leaving no options at all. SEC limped through the year with money given to it by NOAA authorities and other labs.

Figure 8.6: *This main entrance to the Boulder NOAA building was meant to be the back entrance, but because the building was situated near the irrigation ditch, traffic and access had to be routed to the "back." (Will von Dauster, NOAA)*

The long-term resolution of SEC's financial situation required that the FY04 budget return to or surpass 2002 levels. Congress tends to fund projects as it did the year before, with a few minor changes, meaning the FY04 budget amount could end up reflecting the poor FY03 level. Feeling a responsibility to inform users of the possible cuts to SEC services that would result from another disastrous year of poor funding, SEC reported that it planned to deliver services as usual for that year, but might not be able to the next year. The response was tremendous. SEC received countless e-mails and calls from worried and supportive users. A few of the major corporate-user leaders took the issue to their contacts in Washington. Some wrote letters to their members of Congress or to NOAA leaders. Such an outpouring of support gratified the beleaguered SEC staff but did not immediately fix anything.

In August 2003, as Congress considered the FY04 appropriations, a Senate appropriations committee staffer decided that NOAA had strayed in its mission and inserted into his report to the Senate these comments: "Solar Observation—The 'Atmospheric' in NOAA does not extend to the astral. Absolutely no funds are provided for solar observation. Such activities are rightly the bailiwick of [NASA] and [USAF]." In translation, this seemed to mean "no space weather." The report clearly suggested that the USAF or NASA should take on the work done by SEC, and the tables accompanying the report showed zero dollars allotted for SEC. NOAA frowned upon having part of its organization eliminated, so NOAA lawyers went looking for clear legislation that showed that Congress had intended the services and research done by SEL to be a part of NOAA when it was formed in 1970. This organization—which traces its roots back to the 1940s within the DoC, and has played a central role in international committees and cooperation since the 1950s, and had been integral to the NSWP in the 1990s—needed to justify its existence to an uninformed Congress. The lawyers finally found enough evidence that SEC belonged in NOAA, despite the fact that the congressional bill that had created NOAA had never spelled out what exactly the agency would do.

On October 30, 2003, the House Subcommittee on Environment, Technology, and Standards held a hearing titled "What Is Space Weather and Who Should Forecast It?" The SEC and its services sat in the spotlight. The contrast between the testimony given and the zeroed-out appropriation plan could not have been more contradictory.

The speakers who were invited to testify were Dr. Ernest Hildner (director, SEC, NOAA); Colonel Charles L. Benson Jr. (commander, U.S. Air Force Weather Agency, Offutt Air Force Base, Nebraska); Dr. John M. Grunsfeld (chief scientist and astronaut, NASA); John Kappenman (manager of Applied Power Systems, Metatech Corp., formerly of Minnesota Power); Hank Krakowski (vice president of corporate safety, quality assurance, and security, United Airlines, Chicago); and Dr. Robert Hedinger (executive vice president, Loral Skynet, Bedminster, New Jersey). These representatives of some of the most important users of space weather services were prepared to speak about how their industries would be affected by the loss of SEC. The summary testimony from Colonel Benson ran as follows:

> Over the last several decades in which the Air Force and NOAA have analyzed and forecasted the space environment for operational users, we have learned a valuable lesson: space weather is a complex and costly undertaking. Our solution has been to leverage each other's resources, achieving efficiency by concentrating on those things we each do best.

> Our nation is becoming increasingly dependent on space technology. Although the science of space weather is still in its infancy—which some have compared to the meteorological capability of this country in the 1950s—we are on the verge of improved capabilities from new models and data sources that will provide more accurate space weather services. SEC is at the forefront of this movement. The nation's investment in space weather capabilities will yield great future dividends, just as the investment in terrestrial weather fifty years ago is paying off today. The synergy of the two complementary space weather forecast centers at SEC and AFWA has proven to be a national asset to the security and prosperity of the United States. ... We urge this committee to advocate for a healthy and stable SEC so that this critical capability for military and civilian users will continue into the future.

Repeatedly, testifiers emphasized the importance of SEC to their own missions. There was no question of having NASA, or the U.S. Air Force, or even commercial vendors assume the responsibility for space weather; everyone who testified relied on SEC to do their jobs. NASA and the USAF made it clear that they had no interest in taking over this responsibility. In their view, NOAA had played an important part up to now and should continue. NASA scientist Grunsfeld spoke to this issue as he concluded his testimony:

> It is not within NASA's mandate as a research and development agency to provide the operational forecasting services currently provided by the SEC. In addition, the technical capacity, budget and expertise required to perform this activity could not transition to NASA without impacting our other ongoing space flight operations and research.
>
> The NOAA SEC has a unique complement of people, experience, and resources that allows it to provide a high level of service to its space weather customers. There are no other sources, either domestic or foreign, that can provide this type of support. ... The capability to monitor and forecast this environment should remain with the agency that has the mission and the proven expertise to respond to all of these customers.

The subcommittee members were clearly impressed, both by what they heard about the importance of the space weather program being a part of NOAA and by the acclaim given to SEC by testifiers who declared that it was the most responsive and helpful government agency they knew. To impress the subcommittee even further, the Sun did its own testifying. In the days around the hearings, large space weather storms from a surprisingly active Sun wreaked havoc with cell phones, television service, air navigation systems, and electric grids. What later became known as the Halloween Storms set records across the board and continued for nearly two weeks, all unexpected during the declining phase of the solar cycle.

Although impressed with the testimony, the subcommittee could only make recommendations to the House, not change the appropriations bill. Dismaying to all, the enacted funding level for FY04 mirrored that of FY03, so it seemed as though the efforts of many had failed to work. As SEC again received funding at 60 percent late in the first quarter, its services and research were shored up by its parent organization, NOAA's Office of Oceanic and Atmospheric Research (OAR), and by NOAA's National Environmental Satellite, Data, and Information Services (NESDIS).

NOAA's top administrators began planning for SEC to be moved to the National Weather Service (NWS) for FY05, as both groups were service organizations that issued forecasts, warnings, and watches. NOAA submitted the transfer along with its budget, both of which filtered up to become included in the president's budget request for FY05. A last-minute adjustment before the congressional vote on the omnibus FY05 budget finally brought the funding of SEC to almost 90 percent of the FY02 level.*

Now that SEC has become part of NWS, the two organizations are learning about each other's services and needs. NWS staff members know little about the unfamiliar science and services conducted by SEC, though they have been favorably open to the "exotic" new field. Since a low in the 1980s, NWS has fared well in obtaining full funding and, especially important to SEC users, in writing successful requirements for the new weather sensors they need, to be carried on NOAA satellites. SEC needs two instruments not currently flown by NOAA— a solar wind monitor and an imager to watch eruptions from the solar corona; both will improve forecasts for users. NWS plans to help SEC make these forecasting requirements part of NOAA's satellite programs.

---

*As of this writing, the FY06 budget for SEC that was appropriated by Congress and signed by the president, sadly, looked like the FY03 budget— even worse. The budget request, initially submitted by the president, contained $7.2 million for SEC; the House and Senate essentially concurred (the House mark was $7.2 million, the Senate mark $7.0 million). Unfortunately, the conference committee did not reconcile the two but rather settled on $4 million for SEC, a 40 percent cut to the funding originally intended by the House and Senate. Not only were the much-delayed ten new hires cancelled, but the SEC staff could not survive at this funding level. NWS was required to transfer a minimal amount of its funds to keep SEC's federal staff working. SEC can only hope for better times in FY07.

Government employees learn to cope with political and financial upheavals. The staff members who work with and at SEC are deeply devoted to their jobs and usually keep them throughout the funding ups and downs. Top leaders feel these fluctuations most as they wear themselves down dealing with vexing funding issues. Space weather knows no difference between a well-funded year and an ill-funded year—storms continue to rage. As the world steps into the future wedded tightly to its technology, the need for reliable services and products from SEC will increase. Space weather will become part of the common culture and will follow us into the future, wherever we go.

9 The Space Beyond

Everyday experience has shown that the more we learn, the more there is to learn. At one time in human history, we merely knew that the Sun was a mystical source of light. As we have learned more about the Sun—its activity, structure, cycles, and interaction with Earth—we have come to see a complexity that our ancestors would never have guessed. But our modern breakthrough in learning about these aspects of the Sun has opened the doors for virtually infinite further exploration. What lies ahead for us to learn about the Sun and the planets is at least as much as what we currently know. Humans possess curious minds, and people always wonder what is possible. Virtually every solar scientist or forecaster possesses a list of questions about their field that, when answered, leads to another extensive list. Every answer brings us a step further into the future.

Despite billions of dollars invested in space-based observation of the Sun, the ability to observe, predict, and warn of impending solar activity still remains marginal. However, we stand on the cusp of momentous breakthroughs. The near future holds a promise of better forecasts and better understanding of the complex Sun-Earth connection. New aircraft technology will allow ordinary citizens to take flights into space, increasing the need for knowledge about the solar radiation hazard to humans. A robust menu of spacecraft dedicated to space weather observations will improve the quality of forecasts and lengthen the advanced warning time for storms. New models will allow us to begin to explain the linked interactions between objects in the solar system, especially between the Earth and the Sun. A broadening of our comprehension of how space weather affects the solar system will prepare us for venturing into new environments never seen before. With all of these changes, space weather will increasingly become part of the common vocabulary, and the Space Environment Center (SEC) plans to educate the public and help prepare the children who will turn what sometimes seems like science fiction into science fact.

At least three generations of little boys and girls have dreamed about becoming astronauts and blasting into space. Like most dreams, "becoming an astronaut" conjures up what is a tiny fraction of reality: much more of that reality is tedious math lessons, complex physics problems, weightless training, and endless waiting. We dream about looking down from a huge picture window at a serene Earth as we zip toward the rapidly approaching stars in our faster-than-the-speed-of-light spaceship. With some of the fantasy removed, children growing up today will not have to dream much longer. Best of all, spaceflights will not require extensive astronaut training, as for most of us, it will not be the National Aeronautics and Space Administration that takes us into space.

Thanks to the Ansari X-Prize competition, a private company has created a spacecraft for commercial space travel. The X-Prize, founded in 1996, was modeled after the Orteg Prize that Charles Lindbergh won in 1927 by flying solo across the Atlantic Ocean. The $10 million X-Prize was offered to the first private manned spacecraft to exceed an altitude of 100 km (62 miles) twice in fourteen days. Burt Rutan and his Scaled Composites Company designed and built a craft called SpaceShipOne, with the financial help of Paul Allen, cofounder of Microsoft. The two spent over $20 million on this project. The craft's first flight reached an altitude of 103 km (64 miles), passing the required 100-km altitude that is the internationally recognized boundary of space. Within a week, the second flight took off with fewer problems and climbed to an altitude of 115 km (71.5 miles). As the newly minted "astronaut" stepped out onto the runway after the winning flight, the crowds roared in celebration. Burt Rutan, Paul Allen, and Richard Branson (who recently founded Virgin Galactic for space tourism) joined the pilot and his family on the podium for triumphant photographs.

On October 4, 2004, SpaceShipOne received the $10 million X-Prize (Fig. 9.1). Although it was a hefty amount of money, it covered only half of the development costs. However, there were bigger prizes ahead. Beyond winning the prize, Rutan and Allen hoped to start a space tourism industry, which would more than make up for development costs and perhaps lead to further private ventures in space. Rutan and Allen teamed up with Branson, current head of Virgin Atlantic Airlines and Virgin Galactic. The trio struck a deal in which Virgin Galactic would license the SpaceShipOne technology for a fleet of commercial spacecraft. Rutan and others immediately knew the implications of the successful flights. "We do see this as the frontier of transportation around the around the world," said Marion Blakely, administrator of the Federal Aviation Administration. "We know there will be risks, but those risks are worth it."

Among those risks will be space weather. Just like airplane travel, commercial space travel must deal with radiation hazards to passengers and crew, only the risk increases with altitude. As SpaceShipOne flew about ten times higher than a typical commercial airline, the space

Figure 9.1: *On October 4, 2004, the $10 million Ansari X-Prize was awarded to the creators of SpaceShipOne for completing two successful piloted flights into space within two weeks. This picture shows the rocket-powered SpaceShipOne slung below the twin-jet White Knight carrier aircraft. White Knight carried SpaceShipOne to 15 km (9.3 miles) before releasing the spacecraft to rocket to an altitude of over 100 km (62 miles), the accepted beginning of space. (Jim Campbell, Aero-News Network)*

tourism industry will have to watch radiation more closely than the airline industry. Other space weather–caused events, such as single-event upsets, communication failures, and navigation glitches, will no doubt influence the design and operations of future spacecraft.

Satellite technology has progressed in leaps and bounds in the last sixty years—from no satellites to basketball-sized probes in low-Earth orbit to geosynchronous satellites to space-based observatories orbiting at L1. Think of what the *next* sixty years will hold! Scientists have begun to picture future satellite technology, and much of their vision involves advanced orbital mechanics. As previously mentioned, satellites can perch at the Lagrange L1 point because of the balance of forces between the Sun and Earth, and a little nudge in any direction will send the satellite spiraling off into space. The satellite remains between the Earth and the Sun at all times, making this orbit highly favorable for observing the Sun. But what if the satellite at L1 could be moved closer to the Sun and still be kept from orbiting the Sun too slowly or even spiraling into the Sun? The L1 point is a million miles from Earth, while the Sun is ninety-three million miles from Earth; having an instrument closer to the Sun would be extremely useful in getting earlier warnings of storms carried by the solar wind.

Scientists envision satellites propelled by what they call solar sails, which would allow a satellite to orbit anywhere (Fig. 9.2). Considering that space is largely a vacuum, it may seem unlikely that a craft something like a kite could obtain the required lift or drag to keep it in one place. But scientists believe they can use the pressure of light from the Sun to hold the sails aloft. Although the power of sunlight is tremendously small, little friction exists in space to slow down the sail. The sail itself will be made of smooth sheets of thin film coated in a reflective material, creating almost a mirror effect to reflect the sunlight. The light particles will bounce off the mirror-like face, sending the sail in the direction opposite to the particle's rebound. Universities and other

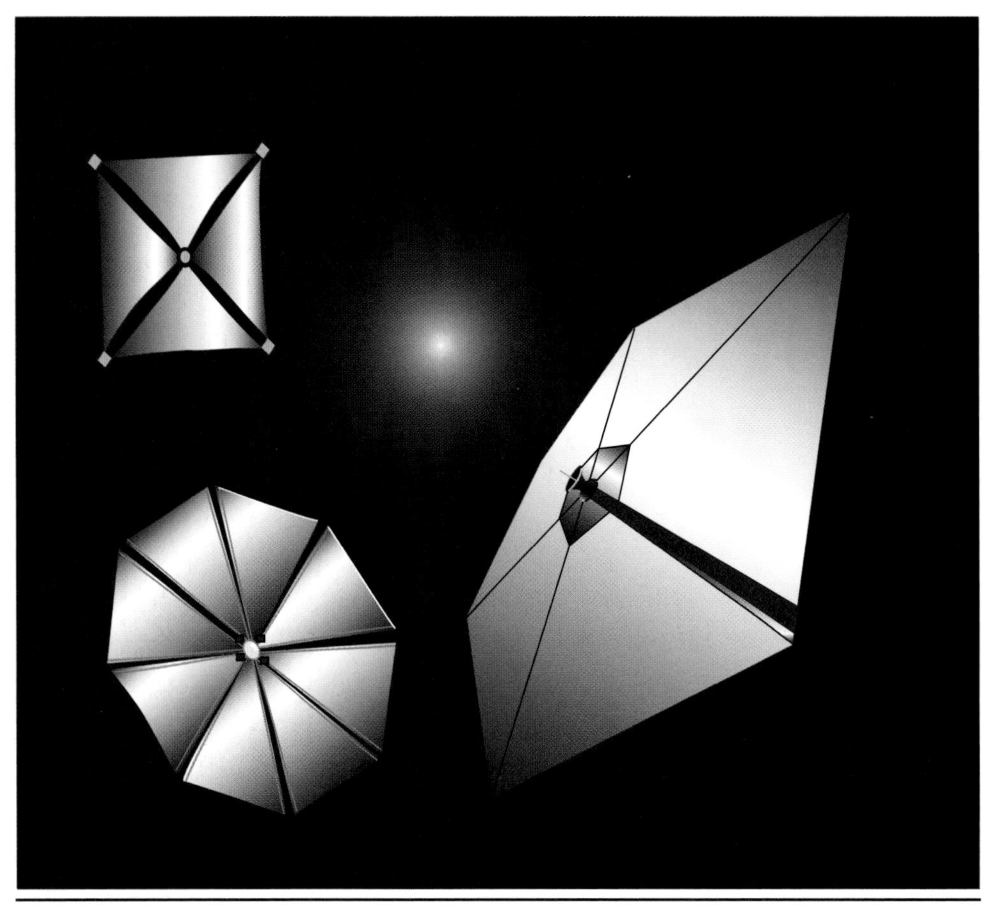

Figure 9.2: *Several concepts for solar sails have been considered by engineers. Sails are generally expected to be made of a thin but tough film, with panels to help "sail" the very large, but light, spacecraft. Solar sails are pushed by sunlight, not the solar wind, whose pressure is only 1 percent that of sunlight. (NOAA)*

research institutions around the country are working on this technology, which will soon be tested in space.

For the near future, rocket scientists continue to hope and plan for new satellites and spaceships. They will play with other special orbits that achieve new goals. Some satellites will assume new orbits very close to the Sun (two-thirds of the way to the Sun from Earth), maybe using a flyby of Venus to kick them into orbit. Satellites with robots will explore other planets, will look at the Sun in pairs or trios, or will observe the far side of Earth's stretched magnetosphere or monitor asteroids. We have gone from the simple beginnings of the Space Race, when simply orbiting the Earth was an enormous achievement, to the more theoretical orbits now planned for the new satellite explorers (Fig. 9.3).

Solar physicists and environmentalists eagerly search for the answer to one question—how much does the Sun influence Earth's climate? The Sun's visible light, of course, drives all of Earth's weather, but this

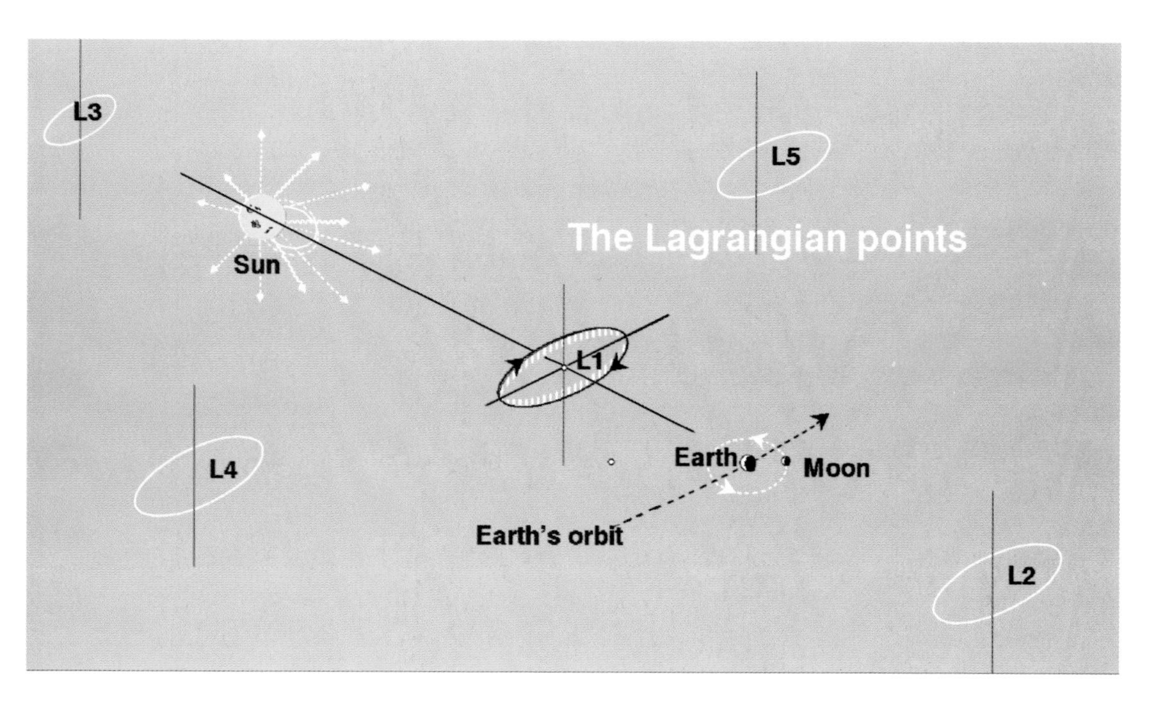

Figure 9.3: *Lagrange (or Lagrangian) points intrigue everyone who has studied Kepler's Laws of motion. Lagrange's work considered the orbits of three bodies (not just two, as Kepler's Laws did). Unconventional orbital mechanics could provide incredible possibilities for future space navigation. (NOAA)*

aspect of the Sun is fairly constant. Clouds, wind, heat—all the components of weather—vary as a result of atmospheric changes, but above the atmosphere the visible light shines on reliably. So how could the constant inflow of light from the Sun cause climatic changes on Earth? Weather is what people experience day to day and describes only the relatively current state of Earth. Climate is what people experience over decades, centuries, or millennia. Scientists have proven that climate can change too; much of the Earth became covered in ice during a massive climate shift during the Ice Age, ten thousand years ago. As seen by the melting of glaciers and the polar ice caps, the world is clearly warmer than it was a hundred years ago.

Scientists have reason to think that the Sun affects Earth's climate. During the last part of the Little Ice Age—lasting about four hundred years—the Sun experienced a particularly quiet time of sunspot activity from 1645 to 1715. Within that quiet time one thirty-year period (called the Maunder Minimum) was particularly anomalous. Astronomers observed only about 50 sunspots, as opposed to a more typical 40,000–50,000 spots in those thirty years (Fig. 9.4). An English solar astronomer, Edward W. Maunder, pieced together the records in 1894 and developed a theory that the unusually cold weather had some relation to the lack of sunspots. Although this discovery failed to impress people during Maunder's lifetime, his theory is now well accepted. Using several sources of data, scientists have identified other quiet periods and corresponding cold cycles and have fortified the belief in the Sun's influence on Earth's climate, championed by scientist John A. Eddy.

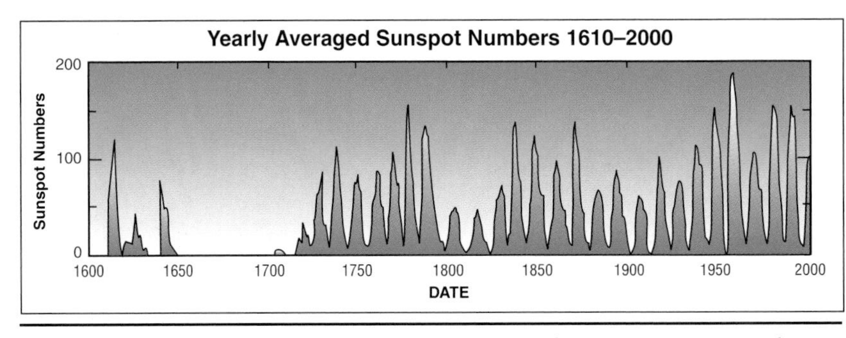

Figure 9.4: *For seventy years, from 1645 until 1715, early astronomers reported almost no sunspot activity—the number of sunspots dropped to virtually none (see flat part of graph). The so-called Maunder Minimum coincided with the coldest part of the Little Ice Age in Europe and North America and has led to theories that the Sun influences Earth's climate. (NOAA)*

Eddy and others have been using the Maunder Minimum to better understand the Sun, but no one is ignoring the cascade of issues that follows the theory. Although the visible light from the Sun is nearly invariable, solar activity variation that leads to space weather comes from 2 percent of the Sun's energy (particles and light outside the visible bands). Is 2 percent enough of a change to cause climate change? The consensus today is yes! Mechanisms to explain how the Sun's variability affects Earth have been debated, but the fact that it does is well accepted.

The immediate importance of this topic lies in the highly contemporary issue of global warming. Scientists know that chlorofluorocarbons created by humans contribute to climatic change. Natural climate fluctuations and the Sun also contribute to such change. But scientists want to know to what extent the Sun's small variability in such spectra as ultraviolet contributes to global warming. The political implications of the answer to this question seem fairly weighty. If the Sun could be credited for a large part of global warming, humans would be absolved of some responsibility, and some would choose to cut back on costly environmental efforts to stop global warming. If, however, the Sun contributes an insignificant amount to the warming, humans would most likely be proved to be the main source.

Another big question on many scientists' lists deals with where space weather ends and terrestrial weather begins. Meteorological scientists have taken an interest in the upper atmosphere at increasingly higher and higher altitudes. At very high altitudes, meteorologists have seen three rare forms of lightning called sprites, jets, and elves (Fig. 9.5). Fast cameras and careful observations show that these lightning-like strikes reach *up* into the upper atmosphere, not *down* to Earth. Although normal lightning originates at altitudes up to 30 km (19 miles), the sprites and jets occur at 60 km (37 miles) or above, an altitude thought to be

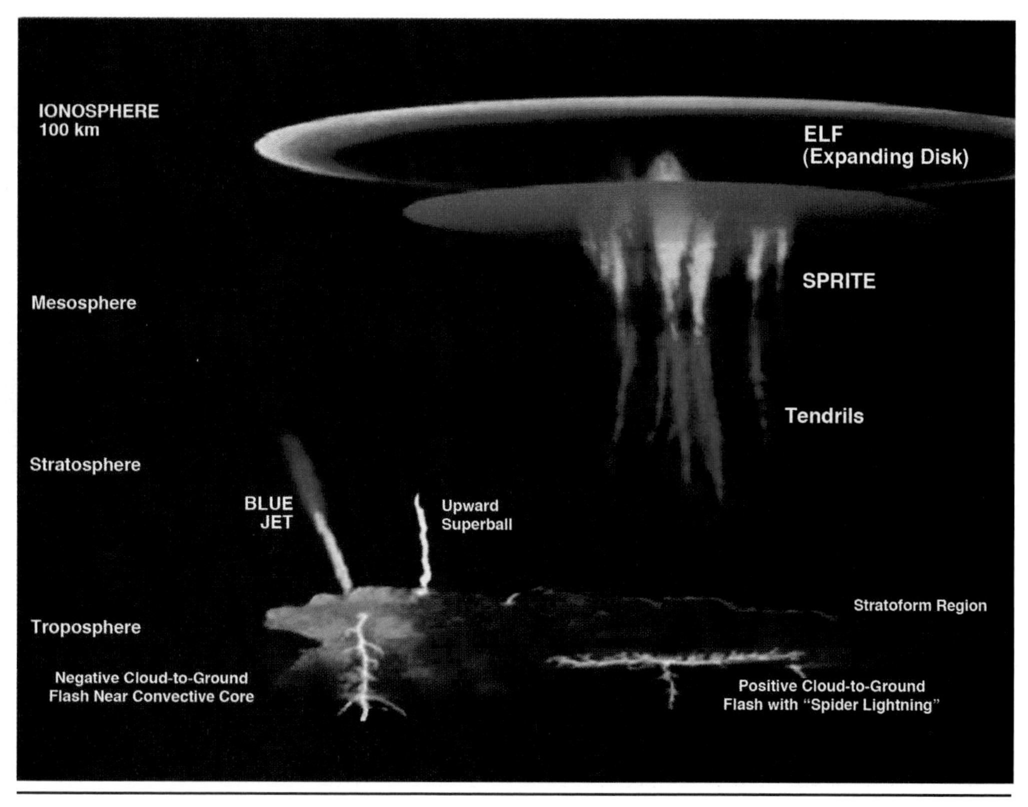

Figure 9.5: *Sprites, jets, and elves occur very high in the atmosphere, leading scientists to wonder if they are terrestrial weather or space weather phenomena. The blue jets extend up from storm clouds into the stratosphere. Red sprites occur in the mesosphere (about 80–90 km [50–56 miles]). The elves appear as expanding ovals in the ionosphere. That these phenomena are seen at all is remarkable, considering their short duration. (www.Sky-Fire.TV)*

above weather. Normal lightning occurs when a huge charge in the clouds discharges to the ground below. What, then, do sprites and jets discharge to if they reach up into space? They seem to be a part of terrestrial weather but reside in the reaches of space. Scientists hope to understand the link between space weather and meteorological weather in the next few decades, if not sooner.

Many cynics claim that the Earth will be destroyed in the next few hundred years and humankind will have to venture to other planets to find a home. Whether this is true or not, space weather will plague the inhabitants of any planet in the solar system, and few other planets have a protective shield like the Earth's (the magnetosphere and atmosphere) (Fig. 9.6). Humans can build shelters to regulate hot and cold, to grow food in, and to protect against radiation, but ultimately space weather will greatly impede life without natural protective shields. Radiation storms and magnetic storms will truly pummel any naked spaceship and equipment, leaving travelers facing catastrophic failures of support systems. We will need to conceive of the new technologies

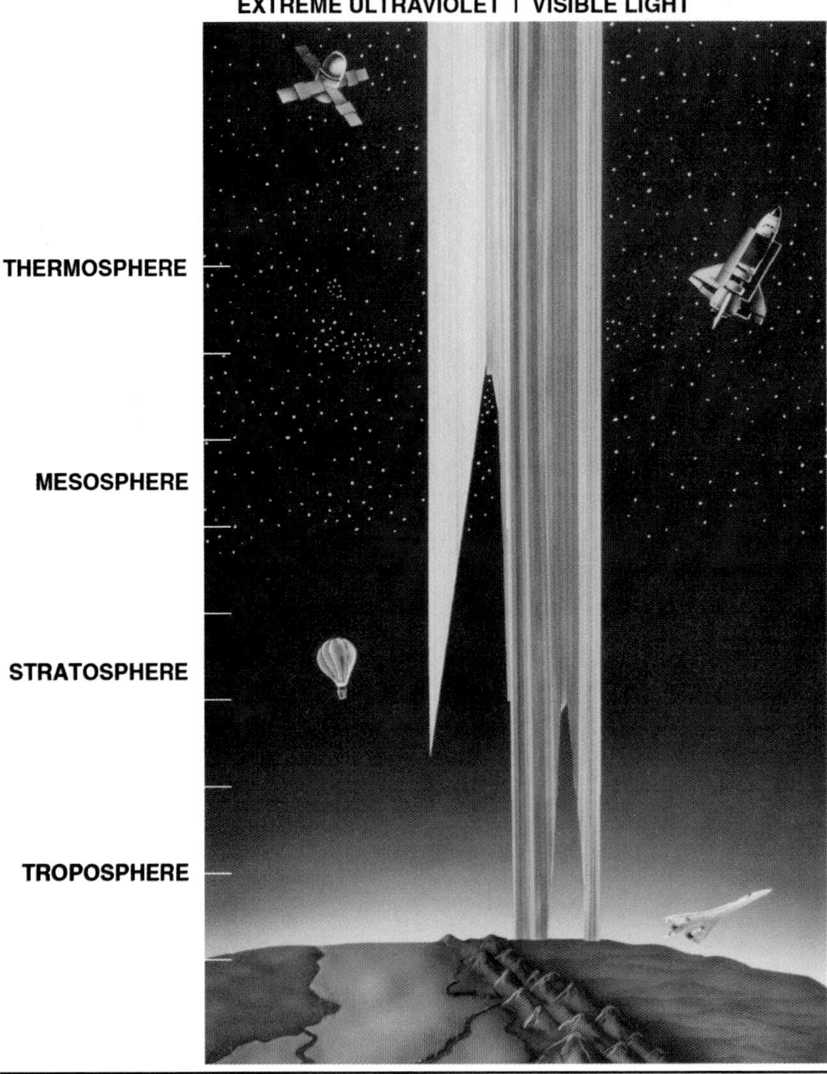

Figure 9.6: *The atmosphere and the magnetosphere stop most harmful rays from the Sun but allow visible light to reach the surface. The boundaries between the different atmospheric levels blend from one to the next and are only approximate. (NOAA)*

that people would need on foreign planets and anticipate the problems with space weather these technologies would encounter. Solar physicists and solar forecasters will be vital in planning long flights to other planets. In considering the state of the Sun, they may decide to recommend that missions fly during Solar Maximum, as the more dangerous cosmic radiation decreases as solar radiation increases.

The issues presented here will no doubt nudge space weather into the common culture. Doctors, sales people, and mechanics will be passengers on spaceflights and will need to understand their radiation risks. Factories, politicians, and environmentalists will fight over global warming and will need to understand the causes of it. Any informed citizen should know about the scientific challenges occurring around him or her. That, of course, means that the public—teachers, students, members of Congress, presidents, engineers, artists—must be educated about space weather (Fig. 9.7). Few people will remain untouched by it.

SEC has a mandate to make the world aware of space weather, its impact on technology, and the challenges it poses to humans. To this end, SEC and its partners have gotten the word out at scientific meetings, teacher conventions, TV science shows, and through various other media

Figure 9.7: *This comic book teaches about space weather in an entertaining and visually delightful way. Created by Zander Cannon, the comic was designed for middle and high school students but rarely fails to attract adults as well. (NOAA)*

outlets. The Internet has revolutionized the sharing of information, but it is invariably the person speaking, giving the tour, appearing on TV, or handing out brochures who provides the real learning. In these settings, personal excitement about the topic rubs off on others. SEC has been remarkably successful at spreading the word about space weather, and much of the public has been suitably impressed.

Perhaps the best classroom for learning about space weather is, in fact, a classroom. Teachers reach thousands of students, students who could engineer the next satellite or create the next model of the Sun, students who will beg their parents to buy them a cell phone or take them into space. The Space Race drew young people into the fields of mathematics and science like never before. Currently the government agencies working in these fields are filled with those students—now approaching their retirement years. The aging of the fields shows the fulfillment of a dream of space travel conceived of in the 1950s. But these will not be the people to lead the way to the next age of discovery. Society must now recruit youngsters into the sciences as effectively as it did in the 1950s (Fig. 9.8).

Many students moving through schools across the nation are ideal candidates for the sciences: they are computer-literate, at ease with technology, and demanding of better and more complex products. Still, many do not know what courses to take, what college to attend, or where they will be able to find jobs. As the fiftieth anniversary of the International Geophysical Year (IGY) approaches in 2007–2008, everyone in the space sciences recognizes that the field stands at the cusp of a new generation. The participants in the IGY (actually three groups made up of the International Polar Year, the International Heliospheric Year, and the Electronic Geophysical Year, along with many partners) will work to bring eager students into this exciting field (Fig. 9.9).

As the sunlight faded into a rich gold and Joe Kunches prepared to leave the forecasting center on a late afternoon in 2005, he remembered when he had first started working at the Space Environment Center. He spun his chair around to look at the forecasting room, which had changed over the years, and remembered forecasting rooms at older facilities. A familiar face passed by in the corridor. Many familiar faces remained from his early days, though many had moved on. The faces of the school groups and the children determined to be rocket scientists and astronauts came more and more often, and he could imagine their faces in his chair, running the forecasting center in the future.

On his way out the door, leaving an operational specialist on duty in the center, Joe saw the stunning iron sculpture of the Sun that adorned the small viewing area in front of the center (Fig. 9-10). The promi-

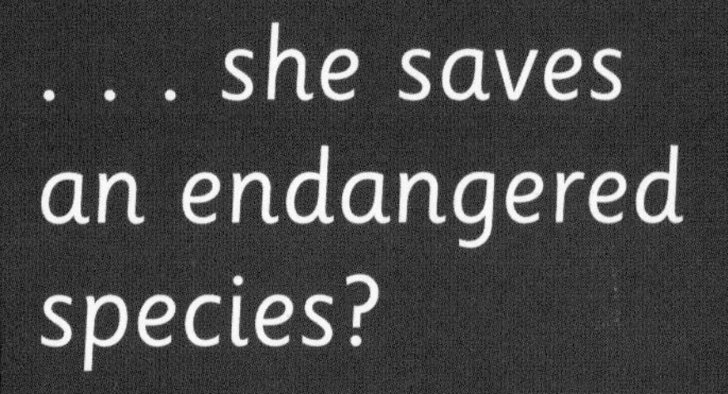

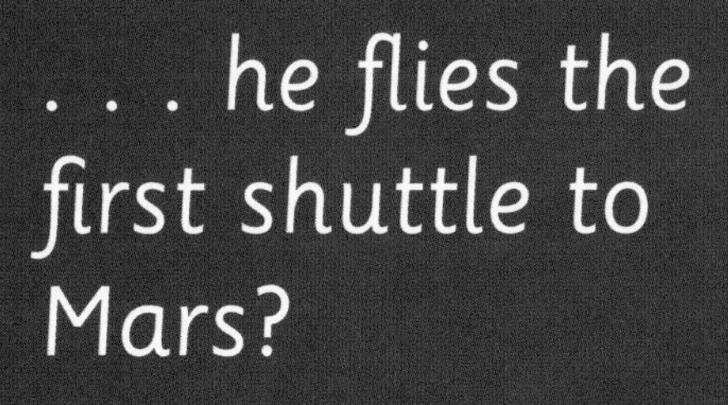

DENVER MUSEUM OF NATURE & SCIENCE
2001 Colorado Boulevard, Denver, Colorado 80205-5798
www.dmns.org or 303.370.6450

Figure 9.8: *Like any science-oriented organization, the Denver Museum of Nature and Science is interested in recruiting young scientists in the making. Aside from being exposed to science, students want to know, "How do I get to be a scientist?" This flyer is part of a campaign to encourage students to follow science as a rewarding, lifelong career. (Denver Museum of Nature and Science)*

Figure 9.9: *Any future trip to Mars, as suggested in this 1991 artist's rendition by Tom Peacock, will involve facing a dangerous Sun. What the exploration vessel will look like is unknown, but perhaps it will be as beautiful as Magellan's flagship. (NOAA)*

nences and flares erupted from the metal surface in a raw, powerful way, and Joe could understand why people worshipped this magnificent celestial body. Down the corridor were hung old pictures of scientists who'd studied the Sun during World War II and the Cold War. Joe especially liked the one of an engineer standing beneath an enormous receiving antenna salvaged from a German World War II antiaircraft system that had a few 50-caliber shell holes in it. He walked by Gary Heckman's old office and remembered the friendly, outgoing man, knowledgeable about the field, dedicated to the users, and full of crazy stories about friends and places around the world.

Joe entered the main lobby, which was guarded by government security staff in these post-9/11 times. The guards waved cheerily to him as he pushed through the revolving doors and stepped out into the fresh Colorado evening. Tomorrow there would be another group of fifth graders to show through the forecasting center, but for now he just breathed in the air and watched the Sun's final sliver of yellow dip below the mountains.

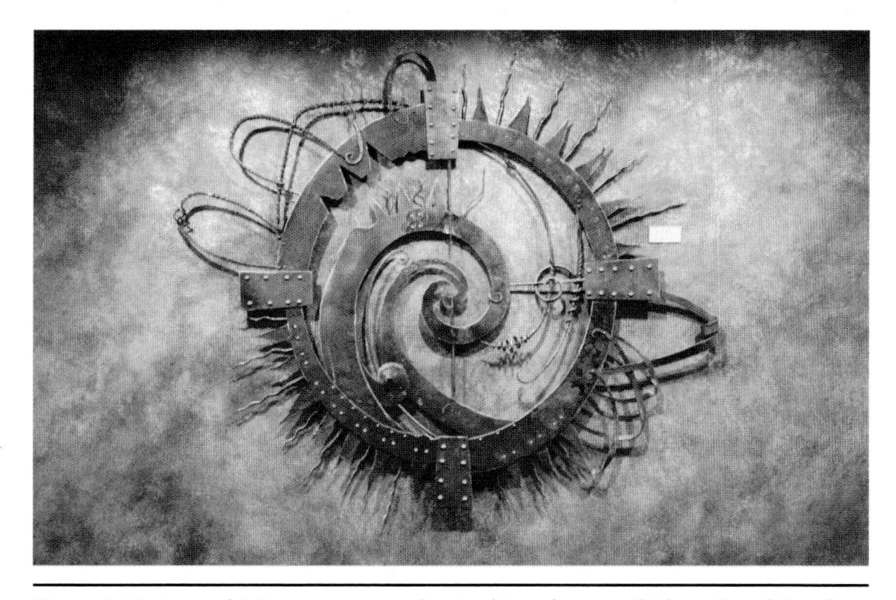

Figure 9.10: *Art and Science come together in this sculpture, which used traditional blacksmithing techniques to visualize the complex features of the Sun. Visitors to the Space Environment Center are surprised to learn so much about our active star, so often seen by them as a plain, yellow ball in the sky. (Black Birch Studio and Forge)*

Supplemental Readings

Akasofu, S.-I. *Aurora Borealis: The Amazing Northern Lights. Alaska Geographic*, 6, 2 (1979) (1994 reprint).

Anderson, Cary, and Dave Parkhurst. *Aurora: A Celebration of the Northern Lights*. Available through http://www.thealaskacollection.com/book.htm. 2004.

Campbell, Wallace H. *Earth Magnetism: A Guided Tour Through Magnetic Fields*. San Diego, CA: Harcourt/Academic Press, 2001.

Carlowicz, Michael, and Ramon E. Lopez. *Storms from the Sun: The Emerging Science of Space Weather.* Washington, DC: The Joseph Henry Press, 2002.

Davies, Kenneth. *Ionospheric Radio*. London: Peter Peregrinus Ltd., 1990.

Eddy, John A., ed. *A New Sun: The Solar Results from Skylab*. NASA SP-402. Washington, DC: National Aeronautics and Space Administration, 1979.

Greenaway, Frank. *Science International: A History of the International Council of Scientific Unions*. Cambridge: Cambridge University Press, 1996.

Harding, R. *Survival in Space*. New York: Routledge, 1989.

Jago, Lucy. *The Northern Lights: The True Story of the Man Who Unlocked the Secrets of the Aurora Borealis*. New York: Alfred A. Knopf, 2001.

Odenwald, Sten. *The 23rd Cycle: Learning to Live with a Stormy Star*. New York: Columbia University Press, 2001.

Websites

NASA sites are excellent and reliable. Use a search engine to find your topic and choose the NASA URLs to pursue.

Rice University Space Weather Resources, space.rice.edu/ISTP/#Education
Space Environment Center, www.sec.noaa.gov
Space Science Institute, www.spaceweathercenter.org
Windows to the Universe, www.windows.ucar.edu

Index

Note: A page locator followed by "f" refers to a figure on that page.